UNI

LARGE PARKS

ays

the d

JULIA CZERNIAK
AND GEORGE HARGREAVES

LARGE

Contributors

JOHN BEARDSLEY

ANITA BERRIZBEITIA

JAMES CORNER

JULIA CZERNIAK

GEORGE HARGREAVES

NINA-MARIE LISTER

ELIZABETH K. MEYER

LINDA POLLAK

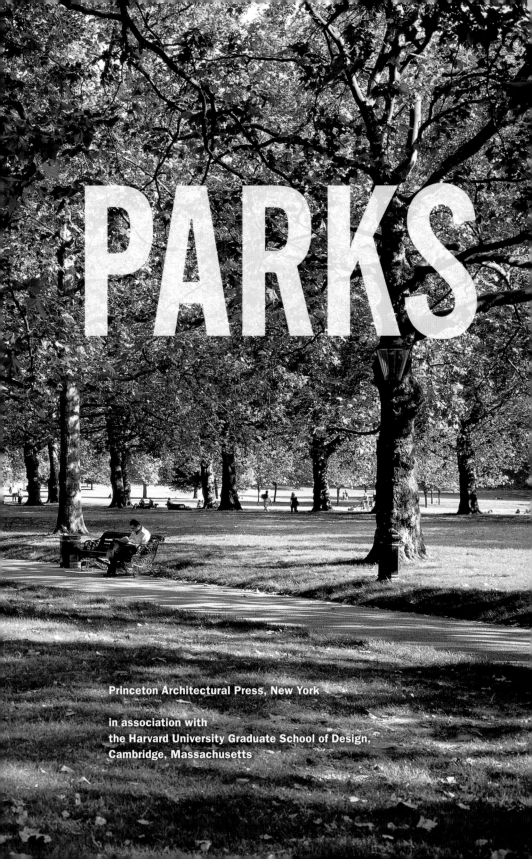

PARKS

Princeton Architectural Press, New York

in association with
the Harvard University Graduate School of Design,
Cambridge, Massachusetts

Published by
Princeton Architectural Press
37 East 7th Street, New York, NY 10003

For a free catalog of books, call 1.800.722.6657.
Visit our web site at www.papress.com.

Editors: Scott Tennent and Lauren Nelson Packard
Design: Paul Wagner

Special thanks to: Nettie Aljian, Sara Bader, Dorothy Ball, Nicola Bednarek,
Janet Behning, Becca Casbon, Penny (Yuen Pik) Chu, Russell Fernandez,
Pete Fitzpatrick, Wendy Fuller, Jan Haux, Clare Jacobson, John King,
Mark Lamster, Nancy Eklund Later, Linda Lee, Katharine Myers, Jennifer
Thompson, Joseph Weston, and Deb Wood of Princeton Architectural Press
—Kevin C. Lippert, publisher

Library of Congress
Cataloging-in-Publication Data
Large parks / Julia Czerniak and George Hargreaves, editors ;
foreword by James Corner.
      p.   cm.
Includes bibliographical references and index.
ISBN 1-56898-624-6 (alk. paper)
1. Parks. 2. Urban landscape architecture. 3. Landscape ecology.
I. Czerniak, Julia. II. Hargreaves, George, 1962– III. Harvard University.
Graduate School of Design.
SB481.L37 2007
363.6'8—dc22

                                        2006036318

Cover Images
front: Bos Park, hill with forest framing space
back: Parc du Sausset, Villepinte, France. View of meadow and bocage

Frontmatter Images
pp. 2–3: Green Park, London, England.
p. 6: Parc du Sausset, Villepinte, France, view of meadow and bocage.
p. 15: Hyde Park, London, rental chairs on slope down to Serpentine Lake.
p. 16–17: Reintroduced bush at Centennial Park, Sydney, Australia.

The sustained interest in landscape over the last two decades is at once remarkable and obvious, given the impact of its study and built work. Publications such as this reflect and generate continued and thoughtful attention. *Large Parks* follows *The Landscape Urbanism Reader* (2006) and *Recovering Landscape* (1999) as the third in a series of publications from Princeton Architectural Press that engage and advance this topic.

The essays, commentary, and graphic material contained within *Large Parks* were developed through a number of events. One of these was a conference held at the Harvard University Graduate School of Design in April 2003, where the topics of the city, ecology, process and place, the public, and site history were used to frame and launch a discussion about the impact and significance of size relative to the planning, design, and management of parks, past and future. Many of the ideas and speculations that resulted from this conference informed the subsequent development of the essays that follow. Along with this conference was an exhibition of large-park case studies by GSD students that provided thoughtful and innovative material that supplemented the discussion. Encouraged by the interest generated by this material, the idea of a publication about large parks was born. The subsequent provocative development of the material—through colloquia, meetings, discussions, and debates—is a testament to the topic's timeliness and relevance for the design disciplines today.

Many people and institutions have contributed to the realization of this book. For providing support for the Large Parks conference and exhibition, we are grateful to the Harvard University Graduate School of Design, the Graham Foundation, and the GSD's Penny White Fund. At Harvard we would like to thank former dean Peter Rowe for his unflagging support, associate dean Patricia Roberts, Niall Kirkwood for his backing as program director, and the administrative assistance of Edna Van Saun and Aimee Taberner. Exhibition coordinator Dan Borelli's work on the exhibition helped inform the development of the material presented here. We also thank Mark Robbins, dean of the school of architecture at Syracuse University, and Alan Altshuler, current dean of the Harvard University Graduate School of Design. Joint institutional funding enabled the color illustrations.

The presentations and commentary of speakers and respondents at the symposium stimulated our work. We thank Victoria Beach, John Beardsley, Anita Berrizbeitia, Joe Brown, Carol Burns, James Corner, Robert France, Robert Garcia, Adriaan Geuze, Kristina Hill, Walter Hood, Dorothée Imbert, Mary Margaret Jones, Niall Kirkwood, Tillman Latz, Nina-Marie Lister, Sébastien Marot, Elizabeth K. Meyer, Elizabeth Mossop, Linda Pollak, Robert Somol, Alan Tate, Marc Treib, Michael Van Valkenburgh, and Bill Wenck.

A large part of this initiative was driven by our desire to disseminate the provocative and original exhibition material developed by Harvard students. We are indebted to Katherine Anderson, Caroline Chen, Gina Ford, Ananda Kantner, Anna Horner, Lara Rose, Emma Kelly, Darren Sears, Jason Siebenmorgen, Rebekah Sturges, and Michael Sweeney for their tireless efforts.

As teachers, we are fortunate to engage with students who have helped us formulate and advance our ideas over the last few years. Among them are Matt Brown, Tyler Caine, David Call, Shivani Gandhi, Mauricio Gomez, John Jedzinak, Bruce Molino, Mick McNutt, Nicholas Rigos, Christine Rounds, Changsoo Sohn, and Kertis Weatherby, some of whose work is included here. We owe the realization of the large park scale studies to Bruce Davison; without his digital and graphic facility, this important study would not have been included.

Particular colleagues have generously given their time and resources to help this project advance. We are grateful to Linda Pollak and Nina-Marie Lister, who provided critical commentary on early versions of this manuscript and spent a cold January weekend at a farmhouse in Ontario poring over what was to become its final form. We acknowledge and thank those in architecture publishing who push projects such as ours toward realization: Melissa Vaughn at Harvard's Graduate School of Design and Scott Tennent at Princeton Architectural Press.

George Hargreaves would like to give special thanks to Mary Margaret Jones for running the operations of Hargreaves Associates and thus giving him the greatest gift of all—time to think. Julia Czerniak would like to thank Mark Linder for being an equally great critic and father. Without his patience, support, and encouragement, this project would never have happened. This book is dedicated to the editors' other life projects: Azriel, Lucja; and Joseph, Kate, and Rebecca.

Broken ice along the shores of Lake Ontario Park, Toronto.

For most of us, large parks conjure up marvelously visceral images of large-scale green open space, replete with forests, allées, bosques, meadows, lawns, lakes, streams, bridges, paths, trails, promenades, and innumerable types of social space, some small and intimate, others grand and monumental. Large parks are extensive landscapes that are integral to the fabric of cities and metropolitan areas, providing diverse, complex, and delightfully engaging outdoor spaces for a broad range of people and constituencies.

In this collection of essays, the numerous references to well-known large parks such as London's Hyde Park, Paris's Bois de Boulogne, New York's Central Park, or Amsterdam's Bos Park capture what for many people are the ultimate virtues of such places—their full experiential range and consolidation of a public's sense of collective identity and outdoor life. Large parks afford a rich array of social activities and interactions that help to forge community, citizenship, and belonging in dense and busy cities. Their scale (defined in this volume as being greater than 500 acres) allows for dramatic exposure to the elements, to weather, geology, open horizons, and thick vegetation, all revealed to the ambulant body in alternating sequences of prospect and refuge—distinctive places for overview and survey woven with more intimate spots of retreat and isolation. These huge experiential reserves are the great outdoor nature theaters of the city, stages for the performance of natural and cyclical time alongside the reveries of social use and event. Large parks are priceless, and those cities that do not have an effectively designed one will always be the poorer.

In addition to their experiential and cultural effects, large parks are also valued for their ecological functions. These vast tracts of land are effective at helping to store and process stormwater, to channel and cool air temperature in the urban core, and to provide habitat for a rich ecology of plant, animal, bird, aquatic, and microbial life. To use an old but still relevant analogy, large parks function as "green lungs," cleaning, refreshing, and enriching life in the metropolis. Their large, contiguous, non-fragmented yet differentiated area is fundamental to their ecological performance, and aspects of wildness are inescapable.

These profound cultural and ecological virtues of large parks will doubtless guarantee their preservation over time and increase demand for new parks in the ever-expanding city. Today, most current large-scale urban design initiatives around the world typically link dense built development alongside or around large swaths of newly designed open space. In part, these open spaces are bargaining chips to compensate for expansive building, but they can also assume designed dimensions of enormous social and ecological value—enhancing and distinguishing the development as a whole.

This demand for large parks is also stimulated by the huge transition around the world from industrial to service economies, creating a vast inventory of large abandoned sites. These sites—old factory and production properties, closed landfills, decommissioned ports and waterfronts, former airfields, and even neighborhoods and sectors of cities where labor has migrated and left empty tracts of towns—lend themselves to being transformed into radically new forms of public parkland and amenity. Needless to say, the challenges to remediating these sites for new public uses are enormous and have led to the advancement of ecological restoration methods within the field, as well as a new regard for emergent management and cultivation techniques. Concurrently, the post-industrial aesthetic—the tough, machinic, strange quality of these sites—has produced an alternative image to the rural pastoral that has dominated public expectations of what parks should look like for the past two centuries. The time for the reinvention of large parks has never been better.

At the same time that large parks provide so much delight, space, and function, they also pose enormous challenges. While expensive to design and build, they are even more expensive over time to operate and manage. In times of fiscal cutback, parks maintenance is first to be cut, and parks can quickly fall into states of disrepair and dereliction. When this happens, parks become the city's backyard, the venue of illicit use, violence, and dumping—the urban wilderness. Parks need stewards, involved constituents, intelligent managers, and fairly healthy budgets if they are to be effectively cultivated for future generations. They also need reinvention, for the parks of yesteryear are practically impossible to recreate.

Whereas the physical, cultural, and experiential delights of large parks are the focus and raison-d'être of their creation, the ecological, operational, and programmatic aspects are less obvious and less understood. And yet in recent years these latter concerns have become of primary importance in the production of large parks. These are specifically important landscape architectural concerns, and they are rightly the main focus of attention for the authors in this collection. Parks after all are not simply natural or found places; they are constructed, built, and cultivated—*designed*. As design is subject to geometrical, material, and organizational invention, large parks do not have to simulate the pastoralism of rural scenery, and neither do they have to embody the beaux-arts formality of axes and planes. Consequently, the design of large parks remains an open question, one that each designer beginning a new project must ask. Criteria of site, program, constituency, and relevant expression drawn from current ideas drive the process until the park finally takes on form, amplifying, dislocating, and juxtaposing the unique attributes of the project. And even then, given the vast scale and timelines involved, a park's physical and spatial form will likely be subject to further revision, management, and change over time.

These latter issues—design, construction, process, technique, concept, and temporality—form the basis of this book, aimed as it is primarily to landscape architects, both academic and professional. And the crux of the debate concerns the relative attention a designer must pay to fixed form versus open-ended process, or more precisely to the ratio and interaction between these two parts of the large-park equation. Added to this equation might be a third interest, of meaning and content. Along these three lines then—fixed form, open process, and meaning—the various authors stake their claims and establish their relative ratios, with some pushing for greater fixity, others for more open-endedness, and some for more (or less) meaning.

Large parks will always exceed singular narratives. They are larger than the designer's will for authorship, they exceed over-regulation and contrivance, and they always evolve into more multifarious (and unpredictable) formations than anyone could have envisaged at the outset. They are complex, dynamic systems. As such, the designer of large parks can only ever set out a highly specified physical base from which more open-ended processes and formations take root. The intelligence of this base foundation—its geometry, material, and organizational dynamics—is the basis around which diversity, growth, and meaningful interactivity may be both instigated and supported over time. If this staged groundwork is too constrained or too complicated or too mannered, it will eventually calcify under the weight of its own construction; if it is too loose or too open or too weak, it will eventually lose any form of legibility and order. The trick is to design a large park framework that is sufficiently robust to lend structure and identity while also having sufficient pliancy and "give" to adapt to changing demands and ecologies over time.

Many successful large parks of the past succeed ingeniously in fulfilling these various equations—think of Hyde Park, Central Park, or Bois de Boulogne. What makes it much more difficult and substantially different today, however, is the *process* by which large parks get made. Large parks are no longer under the purview of kings or single powerful agencies. Instead, large parks today must deal with huge and multifarious constituencies, comprised of many contradictory and opposing parties, often steered by complicated and conservative bureaucracies. Issues of design, form, expression, and process are quickly subjugated by issues of stewardship, maintenance, cost, security, programming, and ad hoc populist politics. These are all important and valid concerns, and the large park designer needs to show continually how the design, form, expression, and process they are setting forth accommodates and exceeds each of these more prosaic yet inescapable issues. If a design cannot demonstrably do this, the result will be the typical bland, populist pastoral pastiche that passes for most "recreational open space" today, with none of the grandeur, theatricality, novelty, or sheer experiential power of real large parks. This is the challenge for landscape architectural design and the broader public imagination in this century, and some of the most critical parameters and strategies for rethinking such a practice are laid out in the pages that follow.

FIGS. 1, 2: Ken Smith Landscape Architect et al., Orange County Great Park, California, bird's-eye view (top) and view of Palm Canyon (bottom).

Julia Czerniak

*The special value of the Central Park to the city of New York will lie,
and even now lies, in its comparable largeness.*
—Frederick Law Olmsted

This book presents essays on an increasingly hard to define landscape type, the park. Two thoughts motivate our work on existing and planned parks internationally. First, initiating a study of parks selected by size cuts across conventional binary categories of classification—historic or contemporary, built or unbuilt, urban or peripheral, competition sponsored or commissioned—and enables review of landscapes not usually considered collectively. The simultaneous consideration of, say, Central Park's circulation strategies (843 acres, 1858) and Field Operations' organizational strategies for the Fresh Kills Landfill (2,200 acres, 2001) provides opportunities to both contextualize and advance thinking about parks, in this case looking specifically at design strategies that give them resilience.[1]

Second, the adjective "large" foregrounds a set of preoccupations in landscape discourse that relate in complex ways, such as ecology, public space, processes, place, site, and the city. Although these aspects of our environment are present in smaller parks, a large park both contains the space that promotes their full interaction and is tangent to a great diversity of urban influences. Given the number of large parks now being speculated about, designed, and planned—such as the Los Angeles River park system as reimagined by Hargreaves Associates (582 acres); the parque del río Manzanares over the M30 in Madrid (679 acres) by landscape architect West 8 and a team of architects; Lake Ontario Park in Toronto (925 acres) by James Corner/Field Operations; North Lincoln Park in Chicago (1,000 acres), envisioned by more than one hundred teams in an ideas competition, including a winning scheme by Clare Lyster, Julie Flohr, and Cecilia Benites; the Orange County Great Park in California (1,316 acres), by a team led by Ken Smith; or Ayalon Park in Tel Aviv (1,976 acres) by Latz + Partner, part of which is a reclaimed landfill—a study of large park design, management, and use is timely and necessary (FIGS. 1–6).

Large

As a qualifier for parks, size has practical and disciplinary consequences, and as sole criterion, the term becomes critical. When Olmsted used the word

FIGS. 3, 4: West 8 et al., parque del río Manzanares, Madrid, view onto gardens with stairs to the river (top) and view of lily ponds at the Puente de Segovia (bottom).
FIGS. 5, 6: Hargreaves Associates et al., Los Angeles State Historic Park, Cornfields, Los Angeles, California, axonometric of urban connectivity (opposite, top) and the Cornfields site in the extended landscape context of Elysian Park and the Los Angeles River (opposite, bottom).

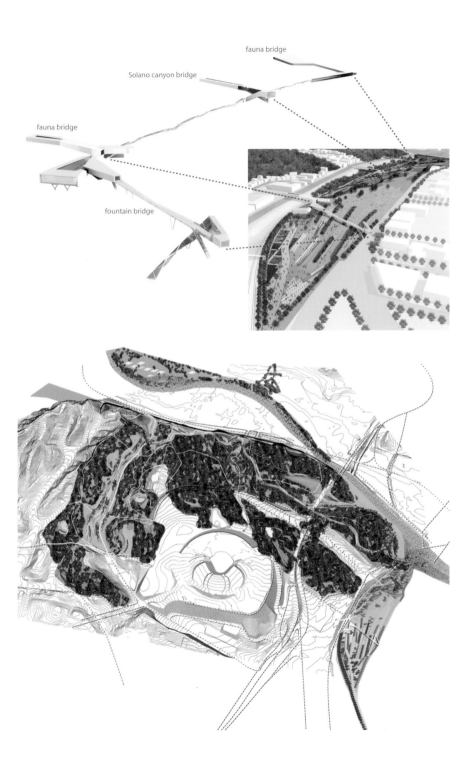

fauna bridge

Solano canyon bridge

fauna bridge

fountain bridge

FIG. 7: Fairmount Park, Philadelphia, location of park extents relative to the city. The park's topographical logic coincides with the Schuykill River watershed.

"park" in his 1870 address "The Justifying Value of a Public Park," he restricted the meaning to "a large tract of land set apart by the public for the enjoy-
ment of rural landscape, as distinguished from a public square, a public garden,
or a promenade, fit only for more urbanized pleasure."[2] In lobbying for a larger
tract of land for Central Park, Andrew Jackson Downing suggested, "Five hundred acres is the smallest area that should be reserved for the future wants of
such a city...in that area there would be space enough to have broad reaches
of park and pleasure-grounds, with a real feeling of the breadth and beauty of
green fields, the perfume and freshness of nature."[3]

In the early stages of America's urbanization, it was easy and rela-
tively inexpensive to acquire the generous space of which Downing spoke.[4]
As a result, many nineteenth-century parks were very large scale. Franklin
Park, part of Boston's Emerald Necklace, extends over 527 acres; the Buffalo
Park System links more than 700 acres; Central Park covers 843 acres; and
Philadelphia's Fairmount Park, a very large park, has grown from 1,061 to
4,411 acres (FIG. 7).[5] Largeness here was necessary to fulfill what was initially understood as the park's primary role: to provide an image of green and the
effects of freshness as an antidote to the industrial city. Although we provisionally take Downing's advice and define "large" as 500 acres, the essays in
this collection expand on both a park's size requirements and the goals of its
design. Indeed, many of the parks discussed in this volume are drawn at the
same scale and compared in figure 8.

Almost one hundred years later, in the context of middle-class flight
from failing cities to the suburbs, urban activist Jane Jacobs viewed largeness
as a liability. In *The Death and Life of Great American Cities*, Jacobs suggested that large parks are especially vulnerable to being "dispirited border vac-
uums," a single massive use of territory that produces danger and possible
stagnation in surrounding urban neighborhoods.[6] She called for large parks
to include "metropolitan attractions," and cited Central Park as a place that
could "make greater use of its perimeter."[7] Bringing uses from deep within
the park to its edge, she posited, could produce "spots of intense and magnetic
border activity," creating a lively connection between the park and city.[8]

Both positions, developed a century apart in vastly different cultural milieus, suggest the question that this collection addresses: For the evolving roles of contemporary large parks, how big is big enough? The motivation
behind size varies, acquired as it was for different needs at different times.
Large amounts of land are indeed necessary to produce the effects of nature,
organize the picturesque, engage adaptive management, design a natural
system, and be economically sustainable—that is, be big enough to include
the resources for the park's own making.[9] Additionally, such large sites,
regardless of the manner and time of their acquisition, enable parks where
the tensions of their formation are evident; the ability to produce complexity exists; new strategies for financing are required; sustainability becomes a
byproduct of economic measures such as lower maintenance and subsequent

FIG. 8: Scale comparison of large parks.

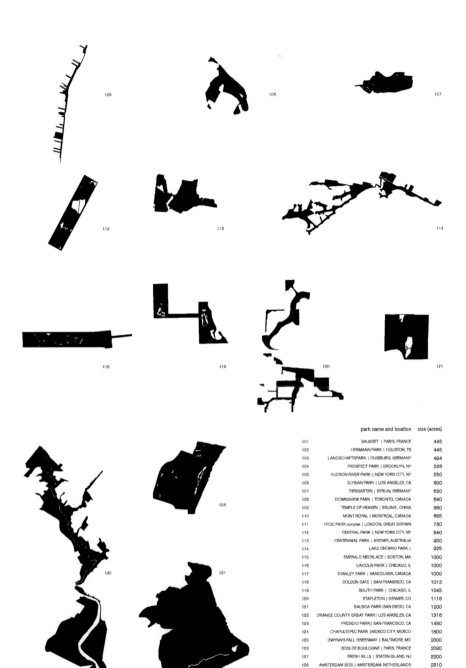

	park name and location	size (acres)
1.01	SALISSET	PARIS, FRANCE	445
1.02	HERMANN PARK	HOUSTON, TX	445
1.03	LANDSCHAFTSPARK	DUISBURG, GERMANY	494
1.04	PROSPECT PARK	BROOKLYN, NY	526
1.05	HUDSON RIVER PARK	NEW YORK CITY, NY	550
1.06	ELYSIAN PARK	LOS ANGELES, CA	600
1.07	TIERGARTEN	BERLIN, GERMANY	630
1.08	DOWNSVIEW PARK	TORONTO, CANADA	640
1.09	TEMPLE OF HEAVEN	BEIJING, CHINA	660
1.10	MONT ROYAL	MONTREAL, CANADA	665
1.11	HYDE PARK complex	LONDON, GREAT BRITAIN	730
1.12	CENTRAL PARK	NEW YORK CITY, NY	840
1.13	CENTENNIAL PARK	SYDNEY, AUSTRALIA	900
1.14	LAKE ONTARIO PARK		925
1.15	EMERALD NECKLACE	BOSTON, MA	1000
1.16	LINCOLN PARK	CHICAGO, IL	1000
1.17	STANLEY PARK	VANCOUVER, CANADA	1000
1.18	GOLDEN GATE	SAN FRANSISCO, CA	1013
1.19	SOUTH PARK	CHICAGO, IL	1045
1.20	STAPLETON	DENVER, CO	1116
1.21	BALBOA PARK	SAN DIEGO, CA	1200
1.22	ORANGE COUNTY GREAT PARK	LOS ANGELES, CA	1316
1.23	PRESIDIO PARK	SAN FRANCISCO, CA	1480
1.24	CHAPULTEPEC PARK	MEXICO CITY, MEXICO	1600
1.25	GWYNN'S FALL GREENWAY	BALTIMORE, MD	2000
1.26	BOIS DE BOULOGNE	PARIS, FRANCE	2090
1.27	FRESH KILLS	STATEN ISLAND, NJ	2200
1.28	AMSTERDAM BOS	AMSTERDAM, NETHERLANDS	2310
1.29	GRIFFITH PARK	LOS ANGELES, CA	4210
1.30	FAIRMOUNT PARK	PHILADELPHIA, PA	4411
1.31	CASA DE CAMPO	MADRID, SPAIN	7000

1 mile
5000 ft   15000 ft

naturalization; and social investment and public input are integral to successful design processes.

Yet size is not the only question, for as important as size is shape, with implications of perimeter and interiority. Parks such as Golden Gate Park in San Francisco (1,013 acres) that emerged in relation to, or even before, the development of the American city have strong figural forms that inform future organizational systems (FIG. 9). For others, such as Stanley Park in Vancouver (1,000 acres), park boundaries coincide with logical natural edges (in this case, a peninsula within the Burrard Inlet). More commonly today, however, designers find themselves making large parks on reclaimed industrial wastelands, brownfields, decommissioned military bases, or landfills whose limits— often political and economic as much as geographic—are imposed, not chosen (FIG. 10). Yet these resultant shapes offer not just challenges but performative advantages. For example, in terms of ecological function, landscape ecologist Richard Forman suggests that the optimum shape for a patch (park) is "generally 'spaceship shaped,' with a rounded core for protection of resources, plus some curvilinear boundaries and a few fingers for species dispersal (FIG. 11)."[10] Design theorist Robert Somol has suggested, with regard to the architecture of OMA, that shape operates with "the graphic immediacy of the logo."[11] One can indeed speculate, with mixed feelings, on a transition from strong figures (nineteenth century) to irregular shapes (twentieth century) to parks as logos (twenty-first century).

In addition to size, the term "large" implies ambition. We are reminded here of Daniel Burnham's call, in the context of planning projects for Chicago, to "make big plans."[12] His corollary encouragement, to "aim high in hope and work" is useful here. "Large" invokes thinking beyond the given, as in Bernard Tschumi's scheme for Downsview Park in Toronto that proposed linkages, beyond the competition boundaries, to adjacent river ravines, promoting the park's ecological role as a wildlife corridor (FIG. 12). Large also implies a considerable amount of energy, vision, commitment, and innovation—by designers, administrators, politicians, and the public they serve—to make these parks happen.

Finally, it is difficult to resist association with Rem Koolhaas's manifesto of "Bigness—the problem of Large," *SMLXL*. Although Koolhaas advances a polemic for architecture, not landscape, appropriating and rephrasing his theorems generates a provocative set of questions to consider when studying, designing, and building contemporary large parks. For example: Can the size of a park alone "embody an ideological program?" Does a large park "instigate the *regime of complexity* that mobilizes the full intelligence of" its discipline? Does size suggest that a large park "can no longer be controlled by a single...gesture?" Can large parks "sustain a promiscuous proliferation of events in a single container?" Do breaks with tradition and composition imply a break with context? Finally, does the large park "need the city," does it "represent the city," "preempt the city," or "is it the city?"[13]

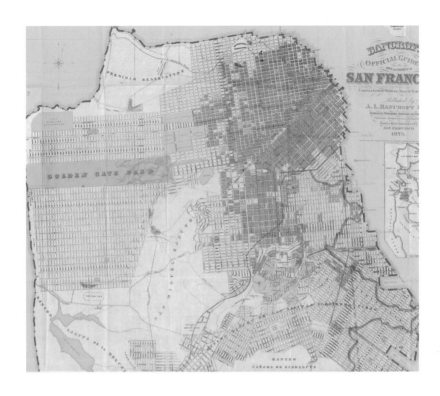

FIG. 9: Map showing area of Golden Gate Park as planned before the city, 1873 (top).
FIG. 10: Shape study, clockwise from right: green system weaving through the redevelopment of Stapleton Airport, Denver; Stanley Park, Vancouver; Golden Gate Park, San Francisco (bottom).

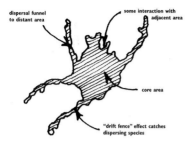

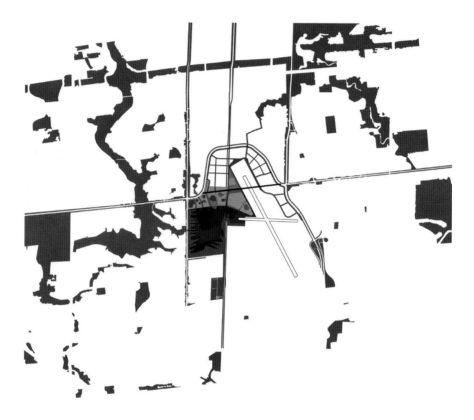

FIG. 11: Ideal patch park shape (top).
FIG. 12: Bernard Tschumi Architects et al., Downsview Park, Toronto, "The Digital and the Coyote" in its extended context (bottom).

Parks

The second term of our title has its own complex history as one of the most debated forms of landscape. In a particularly poetic description in their book on Olmsted, Theodora Kimball and Frederick Law Olmsted, Jr., propose that "whatever the various meanings of the word *park*—to the cottager of Chaucer's time watching the deer over the paling of the manor woods, to the courtier of Louis XIV philandering through the broad allées at Versailles…to the East side urchin of today grasping at his chance for play in Seward Park—it always suggests to us some kind of a green open space with turf and trees."[14] Before the birth of the municipal park movement in Europe in the early nineteenth century, green open spaces were the property of the nobility, to which the public was occasionally invited. The Middle English definition of the word tells us that this land was often enclosed for use as hunting grounds. Frederick Law Olmsted points out that the character of this land, used to support animals, would "be pastoral, with trees sparingly distributed, and having broad stretches of rich greensward."[15]

Yet the character and image of parks, the roles they play, their emergence relative to cities, and their use by various publics has certainly changed. This evolution suggests a considerable amount of both possibility and polemic surrounding park futures. In her book on park design, environmental sociologist Galen Cranz posits that "those with an interest in the character of urban life should seize on parks as one of the vehicles for the realization of their particular visions, and debate around parks should revolve around those visions," and, more specifically, that parks can be "a perfect world in miniature, one that provides norms for the larger world to live up to."[16] Implicit here is a hopeful future for parks, even if what they become no longer resembles the open greenswards of the past. Landscape architect Adriaan Geuze has a different vision, insisting that, in the context of the Dutch landscape, "there is absolutely no need for parks anymore, because all the nineteenth-century problems have been solved and a new type of city has been created. The park and greenery have become worn-out clichés."[17] Geuze appears to be arguing less for the erasure of landscape than for the reconsideration of an exhausted typology. In either case, what is at stake is the appearance and performance of open space within contemporary nature and culture.

It comes as no surprise that Olmsted's parks, work that is as complex as it is misunderstood, appears to be a favorite reference point for designers when pondering these futures. In 1983, Bernard Tschumi set up his park concept for La Villette in Paris as inseparable from the concept of the city, opposing Olmsted's position, or his understanding of it, that "in the park, the city is not supposed to exist."[18] Although La Villette is a small park at 86 acres, this desire to merge the park and the city is present in large parks too. Landscape architect Michael Van Valkenburgh, on the other hand, recently charged to reexamine Olmsted, suggesting there is still much to be honored, studied, and understood from his landscape work.[19]

So "Large" + "Parks" claims a complex conceptual territory. Although the premise that size matters is not complicated, it allows for inquiry at multiple scales and through diverse frameworks that may give rise to new ways of thinking about what parks are, how they look, work, are used and sustained. The parks discussed here span hemispheres and centuries. Beyond size, the criterion for their selection is relevance to urban life. Whether to consider park systems and parklands quickly becomes a question—as contiguous land of varying shape and size is almost the norm rather than the exception of contemporary sites—and not surprisingly some authors refer to them. We initially ruled out all national parks until we found one that was quite urban, Tijuca Park in Rio, participating as it does in the city. Some of these sites have been parks for merely decades, others for more than a century, and some that contribute provocatively to the discourse—such as Hargreaves Associates' competition entry for the Orange County Great Park, James Corner/Field Operations' "Lifescape" for Fresh Kills Landfill, and Bruce Mau's "Tree City" for Downsview Park—have yet to, or will never, be built.

The Essays

The essays collected here are reflections on large parks by some of today's leading landscape architects, architects, design theorists, critics, and historians. They expand on topics at the forefront of landscape discourse as they intersect with large parks.

Landscape ecologist Nina-Marie Lister begins the collection by exploring how recent shifts in ecological thinking—from perceptions of ecosystems as closed, deterministic structures to ones that recognize living systems as open, self-organizing, and unpredictable—affect design and management strategies. Lister's essay, "Sustainable Large Parks: Ecological Design or Designer Ecology?" problematizes these strategies for using complex processes to inform the envisioning and designing of parks. She differentiates between strategies that result in the symbolic and educational—those that represent nature—and those that facilitate the emergence of self-organizing and resilient ecological systems. Although Lister acknowledges that both strategies are valid and desirable, in the context of large parks, she argues, an operational ecology is a basic requirement for long-term sustainability. Lister's essay suggests that the debates surrounding ecology, its nuanced concepts and terms, and its tangencies with the design discipline provide the building block for all subsequent discussions of large parks.

In "Uncertain Parks: Disturbed Sites, Citizens, and Risk Society," landscape historian and theorist Elizabeth K. Meyer follows on this exploration by ruminating on the contemporary large park made in ecologically and culturally disturbed places such as abandoned power plants, obsolete military bases, and landfills whose sites are constituted by debris and toxic byproducts of the city. Pointing out the serious aesthetic limitations and ethical implications of conceiving of the public park as "a space of erasure and amnesia, a timeless pastoral dream floating atop the modern city," Meyer offers instead a view of the park as a space of recollecting and interpreting prior stories of the site. In discussing the formal, spatial, and temporal strategies that designers use to engage these stories, Meyer suggests that there is an audience for and urgency about the topic that goes beyond strategy. It lies in the precise story to be told, which Meyer argues is one that makes legible that the boundaries between toxicity and health, ecology and technology, past and present, city and wild, have long been transgressed. Only then do designers for large parks have the capacity to engage and serve what Meyer calls "the public's perception of ourselves as a collective of citizen-consumers and as residents of risk society."

This preoccupation with thin green veneers that often constitute a park's surface and mask its content is elaborated in "Matrix Landscape: Construction of Identity in the Large Park," by architect Linda Pollak, which focuses on one of the more significant large park sites currently being redesigned in North America, the Fresh Kills Landfill. Taking the size and scale of the Fresh Kills landscape as a starting point, Pollak's essay examines aspects of its complexity that require the development of strategies to engage difference. That is, Pollak argues that the entries for the international design competition held in 2001 for the redevelopment of this site foreground ways to engage the dynamic and heterogeneous aspects of a site—a necessity when conceptualizing, planning, and designing large parks—instead of promoting an illusion of a stable whole. A key device that Pollak culls from the competition entries to develop this approach is a conceptual and representational strategy she calls the "matrix."

Landscape architect George Hargreaves, who uses this strategy in his own practice, draws on his vast experience as a designer, builder, and educator to revisit, both in person and through his students' research (some of which is presented here), seven significant large parks from around the world. This revisiting enables him to speculate on the relationships of these parks, which include the Hyde Park complex in London, Centennial Park in Sydney, and Parc du Sausset on the outskirts of Paris, to their physical sites, which he positions as the fundamental base of any landscape. Touching on issues embedded in large parks such as ecology, human agency, and cultural meaning, Hargreaves describes parks that both resist and engage their sites, noting that site characteristics that could be masked or modified at smaller scales are difficult to disguise across 500 acres. The long-term success of any large park, he argues, depends on the extent to which designers embrace or fight the physical history and systems of a site.

The sensibility of places within parks is reassessed in landscape theorist Anita Berrizbeitia's essay, "Re-placing Process." Here Berrizbeitia examines potential associations between making places of lasting identity and value, and facilitating the natural and cultural processes that transform them. Although the engagement of processes, with their associated temporality, is integral to park design, preoccupations with change must coexist with places that designers envision and owners maintain. Distinguishing between the concepts of "place" and "site," Berrizbeitia proposes a reframing of place that suggests it as at once flexible, emerging, culturally encoded, socially dynamic, and visually powerful. Berrizbeitia leaves us asking, in the context of much of contemporary theory that tries to polarize these concepts: Are place and process really at odds?

In "Conflict and Erosion: The Contemporary Public Life of Large Parks," critic John Beardsley addresses both the multiple, often conflicting publics that use large parks and the possibility of even achieving the historical qualifier of "public." Although Beardsley's observation about the blurring of public and private space is not new, he explores the significant impact that this reality has on the design and management of parks. From the development of public-private partnerships that increasingly plan, finance, maintain, and police parks, to the parcelization of space to accommodate the needs of various user groups, to the increasing appearance of private concessions within park bounds—the erosion of public space and its division among multiple users leaves designers of large parks with many challenges. Beardsley remains convinced, however, that notwithstanding nuanced and even pessimistic views of public space, large parks remain some of the most reliable places for the interaction of the planned and unplanned, sanctioned and unsanctioned—for the unfolding of civic life.

In "Legibility and Resilience," I discuss preoccupations in contemporary park design, as formations of the city shift from dominant centers with dependent peripheries to the polycentric metropolis, from parks that emerged in relation to nineteenth-century industrial cities such as San Francisco to those that appear after the metropolis, such as the 1,000-acre regional park system that weaves through the redevelopment of Stapleton Airport in Denver, an expanding urban mosaic. Using the winning schemes from recent international design competitions as my case studies, I examine how large parks can have significant ecological, social, and generative roles in the contemporary city. These schemes, and many successful large parks of the past, share two essential characteristics: legibility and resilience. That is, they must be understood in their intentions, organizations, and imagery, and they must be capable of experiencing disturbance while maintaining their sensibility and function.

NOTES

1. Some of the issues that preoccupy us were raised more than fifteen years ago at two symposia, both held in 1992, the first in Minneapolis at the Walker Art Center and the second in Rotterdam. These events led to two excellent publications on parks: *The Once and Future Park* (New York: Princeton Architectural Press, 1993) and *Modern Park Design: Recent Trends* (Amsterdam: Thoth, 1993). However, these and many other publications and conferences on parks never specifically address the consequence of size.

2. As cited by Frederick Law Olmsted, Jr., and Theodora Kimball in *Frederick Law Olmsted, Landscape Architect, 1822–1903* (New York: G. P. Putnam's Sons, 1928), 3.

3. Ibid., 27.

4. See Witold Rybczynski's *City Life: Urban Expectations in a New World* (New York: Scribner, 1995) for a thorough discussion on the relationship of city growth and parks in the nineteenth century.

5. Lists such as these appear in other publications, including Rybczynski, *City Life*, 124, and Charles Edward Doell, *A Brief History of Parks and Recreation in the United States* (Chicago: Athletic Institute, 1954), 25.

6. Jane Jacobs, *The Death and Life of Great American Cities* (New York: Random House, 1961), 257.

7. Ibid., 265.

8. Ibid., 266.

9. For example, Downsview Park in Toronto (1999) was conceived as 320 acres of park plus 320 acres of development to economically support the making of the park.

10. Wenche E. Dramstad, James D. Olson, and Richard T. T. Forman, *Landscape Ecology Principals in Landscape Architecture and Land-Use Planning* (Cambridge, MA, and Washington, D.C.: Harvard University Graduate School of Design and Island Press, 1996), 32.

11. Robert E. Somol, *Content* (Cologne: Taschen, 2004), 86–87.

12. As cited by Thomas S. Hines, "No Little Plans: The Achievement of Daniel Burnham," *Museum Studies* 13, no. 2 (1988): 105.

13. Rem Koolhaas, *SMLXL* (New York: Monacelli Press, 1995), 495–516.

14. Olmsted and Kimball, *Frederick Law Olmsted*, 3.

15. Ibid., 211.

16. Galen Cranz, *The Politics of Park Design: A History of Urban Parks in America* (Cambridge, MA: MIT Press, 1982), 246.

17. Adriaan Geuze, "Moving Beyond Darwin," in *Modern Park Design*, 38.

18. As cited by Bernard Tschumi, *Cinegram Folie: Le Parc de La Villette* (New York: Princeton Architectural Press, 1987), 1.

19. Michael Van Valkenburgh, "Burying Olmsted," *Architecture Boston* (March/April 2003): 7.

FIGS. 1, 2: Yorkville Park, Toronto, designer ecology, 2006.

SUSTAINABLE LARGE PARKS:
ECOLOGICAL DESIGN OR DESIGNER ECOLOGY?

NINA-MARIE LISTER

Large parks are complex systems, and as such, parks with an area in excess of 500 acres within contemporary metropolitan regions warrant special consideration and study.[1] In particular, large parks pose specific challenges for long-term sustainability in terms of design, planning, management, and maintenance, principally due to their actual and potential biodiversity coupled with the complexity inherent in their ecology and program. Indeed, "largeness" is a singularly important criterion that demands a different approach to design, planning, management, and maintenance—one that explicitly provides the capacity for resilience in the face of long-term adaptation to change, and thus for ecological, cultural, and economic viability. This chapter explores such an approach to design as a response to issues of complexity and sustainability in the context of "large."

In parks of smaller area in urbanizing landscapes, ecological structures and functions are often significantly altered through habitat fragmentation, reduction and simplification, partial restoration, or even complete re-creation. Such areas usually require intensive management to maintain the ecology in place. Although ecological considerations do play into the design (and its contingent planning and management activities) in smaller parks, I suggest that this is *designer ecology*—an ecology that is vital, indeed essential, for educational, aesthetic, spiritual, and other reasons. Yet this is largely a symbolic gesture provided by such parks' designers to recall or represent nature in some capacity (see, for example, Toronto's Yorkville Park, FIGS. 1, 2). Designer ecology, while valid and desirable in urban contexts for many reasons, is not operational ecology; it does not program, facilitate, or ultimately permit the emergence and evolution of self-organizing, resilient ecological systems—a basic requirement for long-term sustainability.[2]

We ought to appreciate the role of designer ecology in small parks for the reasons stated above, as well as for punctuating and accentuating human agency in landscape. From an operational ecological perspective, however, smaller parks cannot reasonably be self-sustaining, nor thus resilient ecosystems, unless they are functionally connected through robust landscape linkages to other similar areas. Smaller parks typically have simpler programs that are less likely to conflict with ecological goals of conservation and protection. Although smaller parks may have any number of interested stakeholders, design, planning, and management processes continue largely to rely on traditional approaches using discipline-based teams of experts; they are predicated on certainty and control—two characteristics not associated with complex ecological systems.

But large parks are a different matter. Their size, coupled with a diversity and complexity of ecology and program, poses unique challenges for design and specific opportunities for sustainability. For example, they may contain a variety of habitats, some at odds in terms of natural evolution: fast-flowing streams that support trout spawning may eventually become stagnant warm-water ponds if beaver are allowed to do their work. The trout will die out while the beaver flourish. Which is the "correct" state for such a park? If sustainability is the goal, both are valid, but not at once. Design of large parks with conflicting habitats and uses calls for a long-term, bird's-eye view of the whole system, usually by a multidisciplinary team of stakeholders and designers working in collaboration, rather than domination by expertise. Specifically, these parks demand an approach I have generalized as *adaptive ecological design*. Long-term sustainability demands the capacity for resilience—the ability to recover from disturbance, to accommodate change, and to function in a state of health—and therefore, for adaptation.[3] This emerging approach, with some reference to the ecological science on which it is based, is postulated as a response to sustainability for large parks.

Adaptive ecological design is, by definition, sustainable design: long-term survival demands adaptability, which is predicated on resilience. But the discussion of sustainability must not be limited to merely "surviving" in an ecological context. Indeed, one might argue that resilient, adaptive, and thus sustainable ecological design is a fitting metaphor for "thriving," and therefore must include economic health and cultural vitality—two characteristics reflected in contemporary large parks.[4] For example, in contemporary urban areas, escalating land costs coupled with decreasing availability of suitable sites render new parks a costly (and less likely) endeavor. Widespread shrinking of public resources is echoed by demands for public parks to be revenue-generating, thus park planners are under increasing pressure to demonstrate long-term viability and therefore economic sustainability of parks.[5] Compounding these limitations is the demographic reality of the contemporary global city: large parks must be designed for more and different uses by a greater range of users. Thus large parks must be designed for both ecological and programmatic complexity, for both biological and sociocultural diversity, and, accordingly, for all facets of sustainability. Adaptive ecological design is a strategy that moves us toward this goal.

Over the past two decades, there has been a gradual but fundamental shift in the way we understand ecosystems (and thus landscapes) in terms of their structure and function. The perception of ecosystems as closed, hierarchical, stable, and deterministic structures functioning according to a linear model of development has been replaced by the recognition that living systems are open, complex, self-organizing, and subject to sudden but regular periods of dynamic change that are, to some degree, unpredictable.[6]

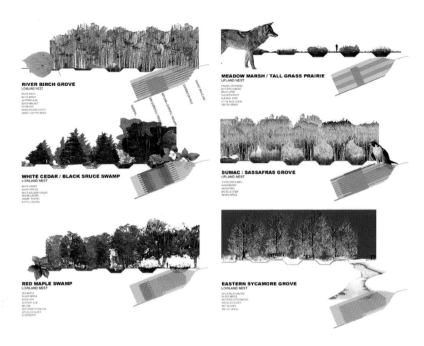

FIG. 3: James Corner + Stan Allen et al., "Emergent Ecologies," Downsview Park, Toronto, diagrams of habitat nests.

The implications of this change in understanding have been variously considered for the planning, design, and management of natural areas and more recently for urban ecosystems.[7]

How might an adaptive, systems-based, ecological design approach be applied to urban and urbanizing ecosystems, or cultural-natural landscapes that characterize this confluence? Despite an emerging discourse in the theory of adaptive management and related literatures—the ecosystem approach,[8] ecosystem health,[9] designed experiments,[10] collaborative environmental planning,[11] etc.—there are few tangible projects. One early prototype of adaptive design in the context of large parks was the design competition held in 2000 for Downsview Park in Toronto. The brief specifically called for an interpretation of ecology consistent with an adaptive, self-organizing, open system, and at least four of the five finalist teams responded with designs that were crafted using language and program resonant with this condition.[12] "Emergent Ecologies," proposed by a team led by James Corner and Stan Allen, depicted an explicitly adaptive plan, which included "seeded evolution" of various habitat "nests," placed in a circuit of both organizational and programmatic ecologies (FIGS. 3, 4). Corner went on to further develop this idea with his team's winning entry for Fresh Kills Park in New York.[13] Yet progress has been slow outside of major design competitions; there has been little substantive exploration of adaptive design, in practice or in empirically supported theory.

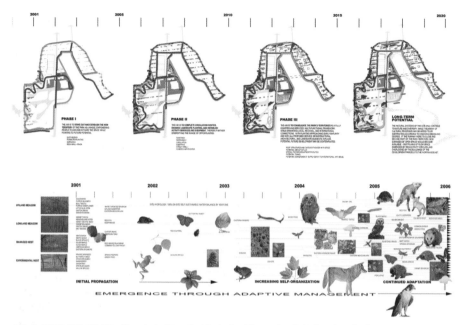

FIG. 4: James Corner + Stan Allen et al., "Emergent Ecologies," Downsview Park, Toronto, adaptive management strategies.

In the context of design, planning, management, and maintenance of large parks, where human recreational needs and creative design goals may conflict with conservation priorities, we must develop complex, layered, flexible, and adaptive design responses. Cities across North America are revitalizing their post-industrial areas, often through the creation of large urban or exurban parks; there is thus an urgent need to consider how designing ecologically, and therefore adaptively, might influence both the art and science of making parks. Furthermore, decision makers must reflect critically on the dynamic intersection of both the behavior of landscape itself and the ways in which designers, planners, and managers work. Such a reflection must necessarily include an analysis of tools, techniques, and strategies for adaptive, ecological design. Within this social dimension of ecological design, there is space to reconsider the role of the designer and his or her interaction with an interdisciplinary and collaborative design team and an interested public. Indeed, the learning potential inherent in the design process in an adaptive context for sustainability may have much to offer as an emergent strategy in large park design, planning, and management.

Ecological Design: A Broader Context for Large-Park Making

Ecological design is an emerging interdisciplinary field of study and practice. In fact, many would argue that it is a transdisciplinary field, concerned with the creation of entirely new applications that may emerge from its progenitor disciplines, or from a synthesis of several. Influenced principally by ecology, the environmental sciences, environmental planning, architecture, and landscape studies, ecological design is one of several rapidly evolving (theoretical and practical) approaches to more sustainable, humane, and environmentally responsible development. It may also be considered a critical approach to navigating the interface between culture and nature. In the broadest sense, ecological design emerges from the dynamic relationship between ecology and decision making. Sim Van der Ryn and Stuart Cowan described ecological design as a hinge that connects culture and nature, allowing humans to adapt and integrate nature's processes with human creations.[14] In modern industrialized societies, human culture and nature are treated as separate realms, yet their interface offers fertile ground for the creation of new, hybridized natural-cultural ecologies and the rehabilitation and rediscovery of others. Ecological design is inspired by the nexus of these worlds and the urgent need to blur the boundary between them; it seizes on the creative tensions between them and may offer insights to "living lightly" with the land. More important, it may also provide a learning framework in which to renegotiate, remediate, and reconsider our relationships to the diverse ecologies that characterize the contemporary urbanizing landscape. (Examples include reclaimed brownfields that, once contaminated, now support heavy-metal-tolerant populations of grasses or recreated wetlands for stormwater management and nature tourism; emerging Great Lakes ecosystems now dominated by introduced species, including salmon that were never expected to survive; or artificial prairie-savannahs, facilitated by urban deer that browse out undergrowth and maintain a tall and groomed tree canopy.) Applied to the current metropolitan landscape, ecological design has particular relevance to large parks and their constituent ecologies and various stewards.

In these ways, ecological design is a vital approach within the broader framework of sustainability. For example, Ann Dale's widely recognized framework is based on the reconciliation of three imperatives: the ecological, the social, and the economic.[15] Three aspects of ecological design may offer opportunities for sustainable design, planning, management, and maintenance of large parks. First, I highlight current understandings from systems ecology that essentially amount to a reconceptualization of ecology and its processes, the result of which demands an adaptive approach to design. Second, I explore the substance of ecological design, characterizing it as adaptive, flexible, integrative, resilient, and responsive in approach, and I reflect on what this means for large parks. Third, I consider that such an approach to design is also a process for engaging the social dimension and may foster the rediscovery of the culture-nature interface—an imperative emphasized in Dale's framework for sustainability.

How is ecological design a relevant context for large parks in the contemporary urban landscape? Ecological design is usually invoked as a means to mimic, model, and even replicate nature's processes and functions—in the work of Laurie Olin, or in Ian McHarg's Staten Island study, for example; it is therefore considered a surrogate model for sustainability.[16] In this sense, ecological design has been associated with "modeling nature"; but this comes with the risk of ecological myopia, in that too much emphasis on strict replication of nature's processes leaves little room for creative synthesis of cultural and natural elements of complex ecologies. Yet there is a far richer interpretation of ecological design, wherein nature is an analog for design, and through such inspired design, a metaphor for human learning. This implies room for a more creative design practice allowing for synthesis with human culture, aesthetics, and ingenuity. And this is critical, reflective space when considering large urban parks.

Outside of several major park design competitions (Downsview, Fresh Kills, and most recently, Orange County), ecological design is principally concerned with the realistic emulation of ecological form, function, and, where possible, process. As an outgrowth of (and to some degree, a fusion between) landscape architecture, ecology, environmental planning, and the building-science aspects of architecture, there is a distinctive functional emphasis in the discipline.[17] Ironically, aesthetics has not been a priority in a discipline that bears the label of "design"; until recently, landscape architecture has been more concerned with applied ecology for reactive remediation—a phenomenon well documented by Corner.[18] The traditional practice of landscape architecture, along with related environmental technologist professions concerned with ecological restoration, have been the progenitors of the new discipline of ecological design, largely as a response to global environmental crises. This is evident in the works of McHarg, Michael Hough, John Tillman Lyle, and others who emphasize that good design should follow the dictates of nature's form and process, often at the expense of creativity and originality.[19] As adaptive ecological design evolves, and as its practitioners seek to define their disciplinary roles, several are beginning to argue fervently for a new creative space for the practice, calling for reconciliation of falsely polarized aspects of art and science, culture and nature.[20]

Despite significant new understandings in ecology over the past twenty years, the field is still largely characterized by a deep schism. As a discipline, the science of ecology is still in polarity: divided between reductionist and holistic perspectives, largely at the expense of a nondualist, integrative systems perspective. This polarity exists in practice and in theory, and is well substantiated in the ecological literature (e.g., the conflict between species and population-scale studies and whole-system studies such as ecosystem energetics).[21] Still, the dominant interpretation and application of modern ecological science is reductionist: decision makers routinely invoke science-based "environmental management," founded on the notion that nature can be counted, measured, and taken apart, a mechanical entity not unlike Newton's outdated notion of the clockwork universe. By extension, conventional wisdom says that nature can to some degree be predicted and controlled, and therefore ultimately managed. But what of more recent insights in systems ecology? What have these to do with design?

Ecological Design is Adaptive Design

Until recently, most ecologists believed that ecosystems follow a linear path of development toward a particular biologically diverse and stable "climax" state. Within the past twenty years, however, research has shown this view to be incomplete.[22] Although ecosystems do generally develop from simple to more complex states, they evolve along any of many possible paths, or even flip suddenly into entirely new states. Ecosystems are self-organizing, open, holistic, cyclic, and dynamic systems, marked by often sudden, unpredictable change. Diversity, complexity, and uncertainty are normal.

It has long been assumed that there is an inherent "balance" or stability in nature, which biological diversity helps to maintain. But this notion of stability is hard to defend in scientific terms. Merely defining what is meant by "stability" is difficult, as living systems experience many fluctuations, such as in weather, populations, and biomass. More recent ecological ideas, based in part on complex systems science, provide a revised perspective of living systems, in which the idea of a single "stable" state is replaced with that of a "shifting steady-state mosaic."[23] In a forest, for example, there are different patches or stands, each of which is a different age. Each patch will grow to maturity, and then fire, windstorm, pest outbreak, or some other disturbance will cause the trees in the patch to die, and growth to start again. Which pieces are at which age changes with time. The patchwork mosaic is shifting constantly over the landscape, even though the landscape remains a forest.[24]

Thus ecosystems have multiple possible operating states and may shift or diverge suddenly from any one of them. In a closed soft maple swamp within a wetland community, for example, changing flows of water can radically alter this state. Extended drought could force a relatively rapid evolution to an upland forest community or grassland. If, in contrast, extended periods of flooding cause high water levels, it would likely become a marsh ecosystem. Red maple (*Acer rubrum*) and silver maple (*Acer sacharrum*) will tolerate floods lasting as long as 30–40 percent of the growing season. Left longer than this, the trees will die, giving way to more water-tolerant herbaceous marsh vegetation.[25] The feedback mechanism that maintains the swamp state is evapotranspiration (i.e., water pumping) by the trees. Too much water overwhelms the pumping capability of the trees, and not enough shuts it down. The current state of the ecosystem is therefore a function of its physical environment coupled with the accidents of its history and the uniqueness of its local context (FIG. 5). Each of these ecosystem states is as ecologically healthy and appropriate as the others, and perhaps more important from a design, planning, or management perspective, there is no single "correct" community for this landscape. This mutability poses both challenges and opportunities for park planners and designers, who have been trained to "choose"

SITE A

SOFT MAPLE FOREST

FLOODING

SOFT MAPLE SWAMP

FORCE

SITE B

CLOSED CANOPY

WIND THROW

BIRTH OF UNDERSTORY

DESIGNED STATE

NATURAL TENDENCY

GOLDEN GATE PARK

PASTORAL

SAND DUNES

FIG. 5: Ecosystem state change in two sites (top).
FIG. 6: Designed state versus natural tendency (bottom).

a future and design for it, with an implicit expectation of permanence. The pressing question is: How should designers respond to this challenge? One strategy may be to anticipate several possible future states, based on the local system history and the social narratives that support it, and to design alternative scenarios that take place temporally as well as spatially. For example, in an ecosystem where localized flooding is a seasonal but not precisely predictable occurrence, park designs can accommodate several ephemeral habitats that appear and disappear based on fluctuating water levels, with minimal management intervention. Indeed, designers might readily embrace the challenges and opportunities posed by the paradox of dynamism: a dance between ephemerality and permanence.

An appreciation of this paradox is important because ecosystems may flip into a new state relatively suddenly. Such flips, properly called bifurcations, have been identified in the Great Lakes, where the dominant ecosystem moves from one characterized by bottom-dwelling (or benthic) species to a deep-water, fish-dominated (or pelagic) state quite quickly, and without warning.[26] Change in an ecosystem as a result of natural catastrophe, such as fire, pest outbreak, or other perturbation, is a normal and usually cyclic event, although it is typically considered catastrophic, even tragic.[27] Consider the forest fires that ravaged Yellowstone National Park, or the windstorms that recently decimated Halifax's Point Pleasant Park or, more famously, Paris's Bois de Boulogne: none of these parks were designed to withstand, let alone accommodate and adapt to, such violent and sudden change. Certainly most large urban parks are characterized by ecological states that are artificially maintained to some degree, in that they require significant inputs (economically and ecologically) to remain in an apparently stable state. Such parks are rarely designed to accommodate either short-term disturbance or long-term, cyclic ecosystem change. In the New World in particular, many of the most prominent and iconic urban parks were designed to emulate Old World landscapes, complete with an introduced ecology that, while comforting and familiar to the colonists, was often functionally at odds with and potentially devastating to native ecosystems. A characteristic example of this phenomenon is Sydney's Centennial Parklands, a system of three large parks covering 890 acres, situated between the city core and the eastern beachfront suburbs. Created as a public open space in 1888 to celebrate Australian Federation, the park was clearly intended to introduce an ecology resonant with the British tradition of grand, formal parks, despite the site's once-dominant but now endangered native *Banksia Scrub* ecology consisting of sand dunes interspersed with coastal swamps.[28] Yet despite this ecological discrepancy, the public perception remains that the area was "transformed from a bleak and sandy area" into the crown jewel of Sydney's greenspace.[29] Similarly, Golden Gate Park in San Francisco is another example of designed ecology that has become problematic and costly to maintain—at least from the perspective of sustainability (fig. 6). While the park is renown for its pastoral landscape of verdant meadows, botanic gardens, arboretum, and lush forests, the dominant ecosystem and native ecology to which much of the park's 1,017 acres would naturally tend is a coastal dune ecosystem. The park is widely appreciated for its deliberate arcadian topography and the variety of habitats and features it incorporates, yet there is a growing awareness of the cost and risks associated with maintaining the created landscape. Still, the public perception is that the park transformed "barren dunes into a forested parkland."[30] In future designs for these and related parks, more emphasis could be placed on ensuring a diversity of plant communities and habitats, with the important inclusion of those that are naturally adapted to normal but cyclic perturbations, including flooding, fire, and wind. Indeed, adaptive (rather than suppressive) strategies are increasingly implemented in parks where seasonal floods or fires are normal, and in some cases necessary, to maintain particular ecosystem types. For example, in Toronto's High Park, annual small-scale controlled burns are undertaken to maintain the oak savannah and prairie ecosystem, deemed culturally significant and ecologically unique to the region.

The ability of ecosystems to recover, reorganize, and adapt in the face of regular change, rather than stability, is critical to their survival. The essence of this primordial ability is resilience. Biological diversity is vital to ecosystems as the basis of resilience, and of the ability of an ecosystem to buffer itself from being pushed into another (potentially less desirable) state, and to regenerate itself following a systemic shift or other disturbance. Biodiversity could be considered analogous to a library of information (some recorded long ago, and some only now being written) that provides not only a wide range of possible pathways for the future development of life but also learned repertoires for responding to environmental change and disturbance.[31]

C. S. Holling's dynamic cycle of ecosystem development is a foundation of the systems view of ecology, which considers ecological organisms and their relationships at multiple scales in time and space (FIG. 7).[32] (Studies include, for example, analyses of ecosystem energetics or the energy flows between trophic levels in a food web.) Living systems evolve discontinuously and intermittently. Following a sudden disturbance, an ecosystem reorganizes to "renew" itself or regenerate to a similar or perhaps different state—one that may be more or less desirable to the humans that inhabit it. Immediately after a disturbance, biodiversity at many scales is critical: the abundance, distribution, and diversity of an ecosystem's structures (e.g., species) and functions (e.g., nutrient cycling) determine its ability to regenerate and reorganize itself, and influence its future pathway.[33] Biodiversity is vital to the normal, healthy functioning of ecosystems because the information it contains and the functions it serves constitute the key elements that determine how an ecosystem will self-organize. In effect, biodiversity forms the palette of future possibilities for an ecosystem.[34]

Most design, planning, and management in an environmental context is based on the assumption that more knowledge leads to certainty, and therefore

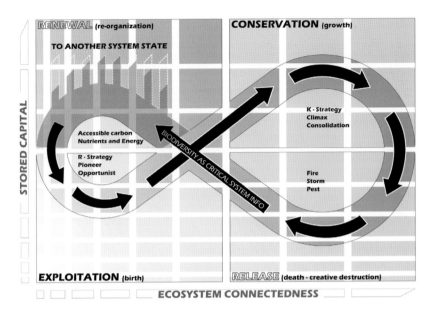

RENEWAL (re-organization)

TO ANOTHER SYSTEM STATE

CONSERVATION (growth)

STORED CAPITAL

Accessible carbon
Nutrients and Energy

R - Strategy
Pioneer
Opportunist

BIODIVERSITY AS CRITICAL SYSTEM INFO

K - Strategy
Climax
Consolidation

Fire
Storm
Pest

EXPLOITATION (birth)

RELEASE (death - creative destruction)

ECOSYSTEM CONNECTEDNESS

FIG. 7: Ecosystem dynamics: C.S. Holling's modified figure eight.

predictability and the success of the design or plan. Although this is resound-ingly true in certain deterministic science and engineering applications, it is not the case with complex living systems. We cannot predict how ecosystems will evolve, change, and behave because they are complex systems that are inherently unpredictable. Of course this does not mean we should fall into the trap of postmodern nihilism and give up trying to design, plan, and man-age; rather, we must accept and embrace change as a normal part of life and, through our designs and plans, adapt to it in a more flexible and responsive manner.[35]

This recent view of ecosystems, and of nature more generally, as open, self-organizing, holistic, dynamic, complex, and uncertain has significant implications for ecological design and other applications in planning and man-agement.[36] We can never determine with precision the consequences of our actions. The current and widely accepted concept of "environmental manage-ment" is an oxymoron, because we can never truly "manage" living systems. Instead, we can refocus our energies on those human activities that provide the context for the self-organizing processes in ecosystems. This implies a profound change in environmental decision making and has concomitant implications for design, planning, and management of ecosystems in general and large parks in particular.

If uncertainty and regular change are inevitable, then we must learn to be flexible and adaptable. Although there is a steadily growing literature on what has been called "adaptive management," there is little empirical and functional understanding of what this means in practice.[37] Given the importance of multiple perspectives at various ecosystem scales (essentially a systems approach), one of the first steps toward flexible, adaptive, and responsive design, planning, or management is to use a diversity of approaches.[38] In general, this means emphasizing small-scale and explicitly experimental approaches that are safe-to-fail, rather than fail-safe.[39] Because ecosystems may change in any number of ways, there may be an infinite number of possibilities for design (and ultimately, management). "Good" ecological design requires a diversity of tools, techniques, and methods. Learning becomes a central goal, leading ideally to continual improvement in design, planning, and management—to long-term adaptation (FIG. 8).

Thus, in developing best practices for ecological design we might consider demonstration projects that emphasize "learning by doing"[40] and "designed experiments."[41] For example, the Huron Park, a medium-sized (325 acres) community cooperative project in Waterloo, Ontario, approved a master plan explicitly designed to accommodate native and non-native species in various and naturally conflicting ecological scenarios, some of which would inevitably disappear, being out-competed by others for nutrients or perhaps management resources.[42] Such projects should be small enough that if they are not successful, they can fail safely, without endangering an entire community, ecosystem, watershed, or habitat. "Failures" or mistakes may provide experiences that can be used in the future. In this way, the "surprising" nature of ecosystems can be turned into a learning opportunity rather than a liability. As Kai Lee observes, "experiments often bring surprises, but if resource management is recognized to be inherently uncertain, the surprises become opportunities to learn rather than failures to predict."[43] I am not aware, however, of any parks planning or management branch that is paid to fail. In the pursuit of accountability, our public agencies consider mistakes or perceived failure a reason to cease funding and remove those in charge—managers and designers who take with them any record of learning, leaving the organization likely to repeat design mistakes rather than learn from them.[44]

Of course good ecological design, such as that required for the long-term sustainability of large parks, must be rooted in rigorous empirically testable science, some appropriately reductionist; it must draw continuously on new knowledge in biology and ecology, among other related disciplines. But adaptive, resilient, and responsive design must also proceed on a broader scale, linked to experience as well as research. Learning through experimentation and action also requires local knowledge for context, as well as field-trained specialists with a range of expertise and research. Fundamentally, adaptive design demands a stronger connection between knowledge and action. "Learning by doing" implies profound changes to our tradition of design, planning, and management, especially in the context of parks. It is still widely assumed that with enough research and knowledge,

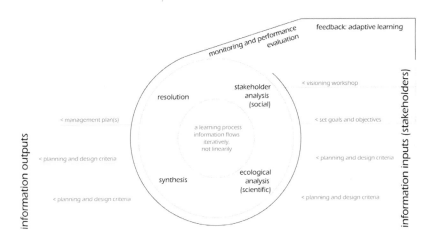

FIG. 8: Adaptive design: a learning process.

ecological systems will somehow be "well-behaved," and that outcomes can be predicted and ultimately controlled. But this is not how the real world works. Adaptation, responsiveness, and flexibility are essential traits, and humans must relearn to live within nature, and perhaps, through design, to reinterpret our relationship with it. If we wish to manage our interactions within nature, we must learn to look to multiple perspectives through a diversity of voices and values, at different scales and in different contexts.[45] To do so in the context of large parks would necessitate meaningful engagement of varied constituents and would force a dramatic reconsideration of parks' values, functions, programs, and ultimately form.

Yet rather than including a diversity of expertise, voices, and professions on our design teams, we tend to favor design and planning processes that are top-down, rigid, homogeneous, and static. The more speculative international design competitions, such as Downsview and Fresh Kills, encourage strongly interdisciplinary teams of artists, scientists, writers, and designers, but the same cannot be said of routine design practice and management structures.

Ecological design, despite the dramatic paradigm shift in ecology, continues to emulate an ecologically deterministic model of nature. As the self-professed "father of ecological design," Ian McHarg defined ecological determinism.[46] His *Design With Nature* was not a suggestion but an imperative, a command to follow the lay of the land in each design. This mandate could be viewed in the broader context of adaptive, resilient, flexible, and responsive design, although his interpretation of ecological "fitness" for good design meant that the correct (truthful) reading of the landscape would necessarily prescribe appropriate design, where form and function are indivisible. His imperative has rarely been interpreted as a call for more open, diverse, or flexible planning and design processes, nor does it typically bring forth a diversity of perspectives, voices, or professions participating on design teams. We can appreciate and understand McHarg's deterministic approach in the context of the 1960s, when science was perceived as a global panacea. But environmental planners should not continue to follow the imperative without some critical reflection on what this means today. The tangled implications of ecosystem complexity, uncertainty, and diversity are significant; these phenomena characterize the dual contexts of large parks and the contemporary urbanizing landscapes—many of which are being reinvented with entirely new ecologies. Well-intentioned but uncritical acceptance of environmental platitudes leaves little room for creative interpretation in the face of ongoing change—and even less for meaningful, adaptive, and responsive design for long-term sustainability of the parks we struggle to make and maintain.

Ecological Design for Engaged Learning

So how do we cope with dynamic change in our urbanizing landscapes and the large parks within these ecosystems? How can we design adaptively, responsive to context and the uniqueness of local conditions, some of which are abused, derelict, or otherwise fundamentally altered, and now far from "natural"? If we can no longer follow McHarg's ecological imperative, certain that "nature will show the way," there must be a new role for humans as creative agents in the process of unfolding, as interpreters of change, as designers once again. As a process of discovery, design implies intentional shaping, manipulation, and (re)creation. In the urban ecological context, it also means recovery of something that has been lost—if not the precise forms of ecologies past, then an attachment to landscape, to nature's rhythms, to place. This process must necessarily be creative and engaging of local people, collaborating in a learning journey based on continual adaptation. Such a process of ecological design might yet move us toward the reconciliation of Dale's imperatives for sustainability.[47]

Sustainability is of course about making choices, in light of necessary limits to growth and a compelling need for equity. As an integral component of sustainability, ecological design incorporates aspects of science and art, culture and nature. Ecological realities should be largely determined through scientific inquiry and learned experience, but in a complex world this knowledge illuminates not "solutions" but choices and trade-offs; decisions are guided by human social choice, by our values. Yet very often in the context of park making in contemporary global cities, there is conflict over these values. Indeed, the sometimes painful process of identifying, revealing, and acknowledging differences in values is essential to achieving a workable design solution. What programs to foster at the expense of others? Which species to protect

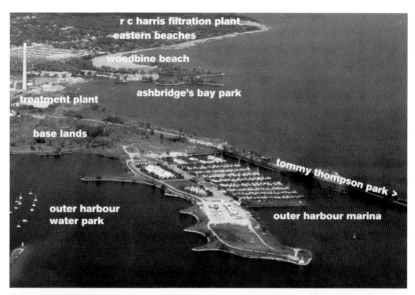

FIG. 9: Lake Ontario Park, Toronto, aerial photo of the eastern end of the site (top).
FIG. 10: Double-crested cormorant colony in defoliated trees on western peninsula of Tommy Thompson Park, adjacent to Lake Ontario Park (bottom).

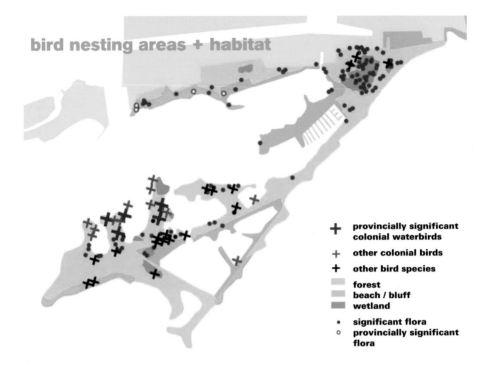

bird nesting areas + habitat

+ **provincially significant
colonial waterbirds**

+ **other colonial birds**

+ **other bird species**

   forest

   beach / bluff

   wetland

• **significant flora**
○ **provincially significant
flora**

FIG. 11: James Corner/Field Operations et al., western end of Lake Ontario Park, Toronto, preliminary site study drawing of bird habitat.

and which to sacrifice? But without the fundamental acknowledgment of difference there can be no respect for diversity, either cultural or ecological, and no reconciliation of Dale's imperatives. And this puts us, as designers, some distance from embracing sustainability.

Yet much of institutionalized planning—the basis for a considerable portion of architectural and landscape design—is still rooted in the science-based deterministic tradition. Ecological science is an essential tool, but when employed without contextual knowledge or social values, science is an insufficient basis for park design. Science-driven bureaucratic approaches nonetheless abound in large parks management and in the implementation of designs and master plans. (Indeed this approach characterizes the National Parks systems in North America.) Planning and design function as top-down, expert-driven, rational activities, relying on management through control. Yet in its social, cultural, economic, and political dimensions, the "nature" of our large parks has very much to do with socially constructed landscape values, and this must be reflected in the design, planning, and management of our parks. Local people should collectively decide which of the many possible futures they want, attainable through choices, trade-offs, trial and error, learning by doing, and flexible management. The designer's role in such a process becomes one of wise facilitator.

Design processes can be potent agents of change. In becoming more open, flexible, and receptive to a diversity of perspectives, and adaptive and responsive to local conditions, they are potentially powerful vehicles for shared, experiential learning by their participants. In several design exercises that have involved a diversity of professions, a range of experts, and meaningful collaboration with local people, I have seen indicators of transformative change among the design team and community members alike. For example, in community meetings leading to Toronto's 925-acre waterfront Lake Ontario Park, bird enthusiasts have fought bitterly with environmentalists who want wind turbines; ecological restorationists want non-native cormorant colonies culled, while others see them as rightful occupants of the park; dog owners, rollerbladers, and joggers have opposed the closing of trails to protect rare plants or breeding birds, while nudists have demanded more clothing-optional beaches. These and other seemingly irreconcilable differences have sometimes been resolved, but more often, simply voiced, heard, understood, and eventually appreciated through facilitated dialog. The public agency overseeing park planning has run a three-year campaign of citizen engagement: public meetings, community workshops, and finally, with the master planning team, design charrettes. Given the complexity of this and other large urban parks, consensus is rarely possible, but compromise is likely. By holding a series of design charrettes in which community members work closely with the design team, social and ecological choices and consequences are articulated, visualized, and prioritized. In many cases, personal and collective changes have emerged from the shared learning experiences (FIGS. 9–11).

In a similar example, the Toronto-based environmental group Evergreen has collaborated with the city of Toronto and the local Conservation Authority to take over the management of a publicly owned industrial and natural heritage site in the center of the city and situated on the banks of the Don River. The Don Valley Brickworks is a 40.7-acre site containing a former brick-making plant, including fifteen heritage buildings and a public ravine and park. The site features a significant geologic formation within the old quarry (considered one of the five most important in North America), industrial buildings (kilns, brick presses, and railyard), and a series of constructed wetlands for stormwater management, habitat protection, and natural heritage enhancement, as well as wildflower meadows, hiking trails, nature interpretation, and a cultural events pavilion. The site is currently being programmed to offer a range of activities and services, from gardening workshops, heritage tours, clay-making, and organic food markets, to a retail nursery, demonstration gardens, and what Evergreen calls "leading-edge green design techniques." According to Geoff Cape, the executive director, the Brickworks is not a park in the traditional sense but a "unique and creative social enterprise that will model sustainability."[48] In 2005, supported through active and creative community partnerships (with some clearly visionary leadership), Evergreen launched their "Rethink Space" campaign to undertake a master plan, completed by Architects Alliance, in which the Brickworks is revealed as a center to connect "culture, nature and community" in the city—and thus manifests the mandate of this nonprofit organization (FIG. 12).[49] Although juxtaposing elements of wild nature with groomed gardens, and arts and cultural activities with the old industrial buildings, the Brickworks does not suffer from conflicting land use goals as one might expect. Rather, the site is engaging creative ecological design as a manifestation of both cultural and natural heritage within the urban context. Not unlike Peter Latz and Partner's Landschaftspark Duisberg-Nord in Germany, the Brickworks site moves the notion of "park" well past nature preservation and into the realm of a learning landscape, designed to teach sustainability by example. Although not a large park in size, the site nevertheless offers a rich palette for both ecological design and designer ecology—both essential strategies of sustainable place-making in the urban condition.

As these examples suggest, ecological design is a useful tool in the learning-based process of park making. In empowering a diversity of voices, values, and participants, this approach may also help to overcome the culture/nature dualism that is a fundamental barrier to sustainability in large parks. This is arguably a potent challenge in the urbanizing landscapes of North America, where layered values—social, ethno-cultural, economic, political, religious, and ecological—collide, split, fuse, and metamorphose. Ecological design holds the potential to navigate the interface of culture and nature in a way that has not yet been part of modern Western history. It may provide

FIG. 12: Evergreen Brickworks, Toronto, 2006.

the intellectual and psychological space to create entirely new, emergent, or hybridized cultural/natural ecologies.

The potential for ecological design to create viable large parks is significant: it lies in the explicit recognition of resilience and adaptation as critical system parameters, and through this, in its ability to elucidate and reconcile the social, ecological, and economic imperatives necessary for long-term sustainability. As a learning process that is adaptive, responsive, and inclusive, ecological design is more broadly insightful in rediscovering, reaffirming, recreating, and reconsidering our place in nature and within contemporary landscapes. The tapestry of the contemporary landscape is complex, woven from many threads, and we need large-scale, responsible ecological design, punctuated by pockets of inspired designer ecologies. The resulting hybrid ecologies of our large parks will likely be at once resonant and dissonant, familiar and unknown, anticipated and unimagined. The sustainable large park, and the landscape mosaic in which it lies, cannot be realized through dialectic argument but rather by creative dialog; it does not serve public space to struggle for either McHargian ecological determinism or postmodern relativism in park design. In learning our way to sustainable design, we must make brave choices. Our designs for large parks must reflect both ecological design and designer ecology, engaged in a relationship of complexity and diversity, and confident in their inevitable uncertainty. This is a key challenge for ecological design and to the successful design and long-term viability of large parks in the contemporary urbanizing landscapes in which we increasingly dwell.

NOTES

I am grateful to James Corner for vision and encouragement, and to Julia Czerniak and Linda Pollak for inspired dialog, provocative collaboration, and insightful comments on an earlier draft. Funding for the background research for this chapter was provided through a Faculty of Community Services' SRC Grant at Ryerson University.

1. "Complex systems" are interconnected networks of processes (or functions) and structures (or elements) whose behavior is generally described as nonlinear, unpredictable, dynamic, and adaptive, and is characterized by the regular emergence of new phenomena and the ability to self-organize. In ecological systems, "complexity" implies a balance between chaos and order within any living system; as such, living systems are said to thrive on the "edge of chaos" or, as Robert Ulanowicz has termed this, "the window of vitality" (conversation with the author, University of Waterloo, 1996); for example, the human body temperature has a very narrow band of optimum performance at 36° C; even a small change in temperature can put the body into a chaotic or disordered state.

2. "Sustainability" here means the inherent balance between social-cultural, economic, and ecological domains that is necessary for humankind's long-term surviving and thriving on the earth. Ann Dale, in *At the Edge: Sustainable Development in the Twenty-first Century* (Vancouver: UBC Press, 2001), has eloquently referred to this balance as a necessary act of "reconciliation" between personal, economic, and ecological imperatives that underlie the primordial natural and cultural capitals on the earth. With this definition, Dale has set the responsibility for sustainability squarely in the domain of human activity, and appropriately removed it from the ultimately impossible realm of managing "the environment" as an object separate from humans—the latter is the conventional implication of "sustainable development." In this chapter, I use the term "management" in the context of Dale's definition of sustainability; that is, in the context of managing human activities within the environment, rather than the environment as object.

3. "Resilience" is used here in the ecological context, as a term developed by the Canadian ecologist C. S. (Buzz) Holling at the University of British Columbia in the mid-1980s. In the general ecological sense, resilience refers to the ability of an ecosystem to withstand and absorb to some degree the effects of change, and following these change events, return to a recognizable steady state (or states). These change events (which Holling has referred to in the vernacular as "surprises"), while usually normal ecosystem dynamics, are also unpredictable, in that they cause sudden disruption to a system: for example, forest fires, floods, pest outbreaks, etc. More specifically, resilience can also mean the rate at which an ecosystem returns to a single steady or routinely cyclic state following a sudden change. The ability of the system to withstand sudden change assumes that behavior of a system remains within the stable domain that contains this steady state in the first place. However, when an ecosystem shifts from one stability domain to another (called "reorganization" via a "bifurcation" or "flip" in system states), a more specific measure of ecosystem dynamics is needed: that of "ecological resilience," which in this context is a measure of the amount of change or disruption that is required to move a system from one state to another, and thus, to a different state being maintained by a different set of functions and structures than the former. Holling's work in resilience has been instrumental to ecosystem managers and ecologists alike in exploring the paradoxes inherent within living systems—the tensions between stability and perturbation, constancy and change, predictability and unpredictability—and the implications of these for management. For a summary account of Holling's work on resilience, see Lance Gunderson and C. S. Holling, eds., *Panarchy: Understanding Transformations in Human and Natural Systems* (Washington, D.C.: Island Press, 2002).

4. See also Julia Czerniak, "Legibility and Resilience," in this volume.

5. See also John Beardsley, "Conflict and Erosion: The Contemporary Public Life of Large Parks," in this volume.

6. See, for example, C. S. Holling, "The Resilience of Terrestrial Ecosystems: Local Surprise and Global Change," in William Clark and R. Ted Munn, eds., *Sustainable Development of the Biosphere* (Cambridge: Cambridge University Press, 1986), 292–320; Gunderson and Holling, eds., *Panarchy*; and David Waltner-Toews, James Kay, and Nina-Marie Lister, eds., *The Ecosystem Approach: Complexity, Uncertainty, and Managing for Sustainability* (New York: Columbia University Press, in press).

7. For the former, see Gary K. Meffe, et al., *Ecosystem Management: Adaptive Community-Based Conservation* (Washington, D.C.: Island Press, 2002); for the latter, see Alexander Felson and Steward T.A. Pickett, "Designed Experiments: New Approaches to Studying Urban Ecosystems," *Frontiers in Ecology and the Environment* 3, no. 10 (2005): 549–56.

8. Waltner-Toews et al., eds., *The Ecosystem Approach*.

9. Ibid.

10. Felson and Pickett, "Designed Experiments."

11. John Randolph, *Environmental Land Use Planning and Management* (Washington, D.C.: Island Press, 2004).

12. Discussed in detail by Julia Czerniak, ed., *Downsview Park Toronto* (Munich and Cambridge, MA: Prestel and the Harvard University Graduate School of Design, 2002).

13. See Linda Pollak, "Matrix Landscape: Construction of Identity in the Large Park," and Czerniak, "Legibility and Resilience," both in this volume.

14. Sim Van der Ryn and Stuart Cowan, *Ecological Design* (Washington, D.C.: Island Press, 1996), 201.

15. Dale, *At the Edge.*

16. See for example Janine Benyus, *Biomimicry: Innovation Inspired by Nature* (New York: William Morrow, 1997), 308; and Ian McHarg and Fritz Steiner, eds., *To Heal the Earth: The Selected Writings of Ian McHarg* (Washington, D.C.: Island Press, 1998), 381.

17. See, for example, Timothy Beatley, *Green Urbanism: Learning from European Cities* (Washington, D.C.: Island Press, 2000), 491; William B. Honachefsky, *Ecologically Based Municipal Land Use Planning* (New York: Lewis, 1999), 256; Frederick Steiner, *The Living Landscape: An Ecological Approach to Landscape Planning* (New York: McGraw-Hill, 1991), 356; and Fred Stitt, ed., *Ecological Design Handbook: Sustainable Strategies for Architecture, Landscape Architecture, Interior Design, and Planning* (New York: McGraw-Hill, 1999), 467.

18. James Corner, "Ecology and Landscape as Agents of Creativity," in George Thompson and Frederick Steiner, eds., *Ecological Design and Planning* (New York: Wiley, 1997), 81–108, and James Corner, "Recovering Landscape as a Critical Cultural Practice," in James Corner, ed., *Recovering Landscape: Essays in Contemporary Landscape Architecture* (New York: Princeton Architectural Press, 1999), 287.

19. See Ian McHarg, *Design with Nature* (Garden City, NY: Natural History Press, 1969), 198; Michael Hough, *Cities and Natural Processes* (New York: Routledge, 1995); and John Tillman Lyle, *Design for Human Ecosystems: Landscape, Land Use, and Natural Resources* (Washington, D.C.: Island Press, 1999), 279.

20. See, for example, Louise Mozingo, "The Aesthetics of Ecological Design: Seeing Science as Culture," *Landscape Journal* 16, no. 1 (1997): 46–59; Corner, "Ecology and Landscape as Agents of Creativity" and "Recovering Landscape as a Critical Cultural Practice"; Charles Mann, "Three Trees," *Harvard Design Magazine* 10 (2000): 31–35; and more recently, Mohsen Mostafavi and Ciro Najle, eds., *Landscape Urbanism: A Manual for the Machinic Landscape* (London: Architectural Association, 2003), and Charles Waldheim, ed., *The Landscape Urbanism Reader* (New York: Princeton Architectural Press, 2006).

21. Explored in detail in Nina-Marie Lister, "A Systems Approach to Biodiversity Conservation Planning," *Environmental Monitoring and Assessment* 49, no. 2/3 (1998): 123–55.

22. See, for example, Holling, "The Resilience of Terrestrial Ecosystems"; Carl Walters, *Adaptive Management of Renewable Resources* (New York: Macmillan, 1986); C. S. Holling, David Schindler, Brian Walker, and Jonathan Roughgarden, "Biodiversity in the Functioning of Ecosystems: An Ecological Primer and Synthesis," in Charles Perrings et al., eds., *Biodiversity Loss* (Cambridge, MA: Cambridge University Press, 1995), 44–83; James Kay, "A Non-equilibrium Thermodynamic Framework for Discussing Ecosystem Integrity," *Environmental Management* 15, no. 4 (1991): 483–95; and James Kay and Eric Schneider, "Embracing Complexity: The Challenge of the Ecosystem Approach," *Alternatives Journal* 20, no. 3 (1994): 32–38.

23. Holling, "The Resilience of Terrestrial Ecosystems"; and Walters, *Adaptive Management.*

24. Frederick H. Bormann and Gene E. Likens, *Patterns and Process in a Forested Ecosystem* (Berlin: Springer-Verlag, 1979).

25. Nina-Marie Lister and James Kay, "Celebrating Diversity: Adaptive Planning and Biodiversity Conservation," in S. Bocking, ed., *Biodiversity in Canada: Ecology, Ideas, and Action* (Toronto: Broadview Press, 2000), 189–218.

26. James Kay, Henry Regier, Michelle Boyle, and George Francis, "An Ecosystem Approach for Sustainability: Addressing the Challenge of Complexity," *Futures* 31 (1999): 721–42.

27. Holling, "The Resilience of Terrestrial Ecosystems."

28. Prior to European contact, much of the eastern Sydney peninsula was vegetated with Eastern Suburbs Banskia Scrub species (ESBS), which are primarily comprised of various banksias, heath-type shrubs, grasses, and sedges, typically found on older, deep sandy soils in the eastern suburbs of Sydney, southward to Botany Bay. Today less than one percent of the original ESBS survives in isolated remnants, including four identified remnants within the Centennial Parklands—all of which have been modified over time by the introduction of non-native trees and grasses. The ESBS ecology has been formally listed as an Endangered Ecological Community in Australia. See Centennial Parklands, http://www.cp.nsw.gov.au/data/assets/pdf_file/718/remnant_bushland.pdf (accessed Oct. 13, 2006).

29. "Centennial Park," travelpromote.com.au, http://discoversydney.com.au/parks/centennial.html (accessed Oct. 13, 2006).

30. Nick Wirz, "Golden Gate Park," Project for Public Spaces, http://www.pps.org/great_public_spaces/one?public_place_id=74 (accessed Oct. 13, 2006).

31. Lister, "A Systems Approach to Biodiversity Conservation Planning."

32. Holling, "The Resilience of Terrestrial Ecosystems."

33. Holling et al., "Biodiversity in the Functioning of Ecosystems."

34. Lister, "A Systems Approach to Biodiversity Conservation Planning"; and Waltner-Toews et al., eds., *The Ecosystem Approach.*

35. Kay and Schnieder, "Embracing Complexity"; Lister and Kay, "Celebrating Diversity."

36. See, for example, Bruce Mitchell, ed., *Resource and Environmental Management in Canada*, 4th ed. (Toronto: Oxford University Press, 2004); Waltner-Toews et al., eds., *The Ecosystem Approach.*

37. See, for example, Gunderson and Holling, eds., *Panarchy*; Holling, "The Resilience of Terrestrial Ecosystems"; Carl Walters and C. S. Holling, "Large-Scale Management Experiments and Learning by Doing," *Ecology* 71, no. 6, (1990): 2060–68; Kai Lee, *Compass and Gyroscope: Integrating Science and Politics for the Environment* (Washington, D.C.: Island Press,1993), 243; Meffe et al., *Ecosystem Management*; Walters, *Adaptive Management*; Waltner-Toews et al., eds., *The Ecosystem Approach.*

38. R. Edward Grumbine, "Reflections on 'What Is Ecosystem Management?'" *Conservation Biology* 11, no. 1 (1997): 41–47; Carl Folke, Thomas Hahn, Per Olsson, and Jon Norberg, "Adaptive Governance of Social-Ecological Systems," *Annual Review of Environment and Resources* 30 (2005): 441–73.

39. Lee, *Compass and Gyroscope*; Felson and Pickett, "Designed Experiments."

40. Lee, *Compass and Gyroscope.*

41. Felson and Pickett, "Designed Experiments."

42. Lister and Kay, "Celebrating Diversity."

43. Lee, *Compass and Gyroscope*, 56.

44. This phenomenon is well documented in business and organizational literature, for example, in Peter Senge's *Fifth Discipline: The Art and Practice of the Learning Organization* (New York: Doubleday, 1990), and is growing in the environmental management literature. See, for example, Francis Westley, "Governing Design: The Management of Social Systems and Ecosystems Management," in Lance Gunderson et al., eds., *Barriers and Bridges to the Renewal of Ecosystems and Institutions* (New York: Columbia University Press, 1995), 391–427.

45. Jim Woodhill and Niels G. Röling, "The Second Wing of the Eagle: The Human Dimension in Learning Our Way to More Sustainable Futures," in Niels G. Röling and M. A. E. Wagemaker, eds., *Facilitating Sustainable Agriculture: Participatory Learning and Adaptive Management in Times of Environmental Uncertainty* (Cambridge: Cambridge University Press, 1998), 47–71.

46. McHarg, *Design with Nature.*

47. Dale, *At the Edge.*

48. Geoff Cape, conversation with the author, 2006.

49. Evergreen, "Evergreen at the Brickworks: Final Master Plan," June 2006, http://www.evergreen.ca/en/brickworks/.

FIG. 1: This red basin within Calloway Copper Mine is connected to a subterranean diversion tunnel outlet, one of the numerous pits and excavations within a 52-square-mile dead zone in Ducktown, Tennessee. When the water surface is agitated, it becomes surreal, turning from blue to deep red, and revealing its low PH and acidic nature.

UNCERTAIN PARKS:

DISTURBED SITES, CITIZENS, AND RISK SOCIETY

Elizabeth K. Meyer

What are we to make of contemporary parks that are planned or built on disturbed sites? Two centuries ago, large parks were created out of former royal gardens and hunting grounds. A century ago, they were located on large rural parcels, on the periphery of expanding cities. Most of those sites were in rural use; a few had prior industrial uses, such as the quarry upon which Paris's Parc des Buttes Chaumont was built in the 1860s. Today they will often be located on the only lands available in metropolitan areas: abandoned or obsolete (and often polluted) industrial lands such as quarries, water-treatment facilities, power-generation plants, factories, steel mills, landfills, military bases, and airports.

I have chosen to use the term "disturbed site" to describe a broad category of polluted or contaminated landscapes previously used for industrial purposes, regardless of whether their pollution was intended or not, acknowledged or hidden, regulated or unknown. These are popularly called brownfields and gray fields; legally known as Environmental Protection Agency–designated Superfund sites; and professionally coined manufactured sites, wastelands, or toxic sites (FIGS. 1, 2).[1] "Disturbed" captures the effect as well as the character of these sites. They have been disturbed by new processes—interrupted and interfered with—and that alteration disturbs us, makes us uneasy, anxious, worried, and agitated. This term also resonates with contemporary theories of ecological science that recognize the significance of disturbance regimes within successional processes and ecosystem dynamics.[2]

These parks vary from renowned built examples, such as Peter Latz and Partner's 568-acre Landschaftspark Duisburg-Nord on the site of the Thyssen-Meiderich blast furnaces and slag heaps once connected via railroad to a network of mills and factories in Germany's industrial Ruhr River valley; to Field Operations' widely published winning entry for Staten Island's Fresh Kills Park, located on a 2,200-acre landfill comprised of a half-century of New York City's garbage; or the recently commissioned Orange County Great Park (awarded to the multidisciplinary team of Ken Smith, Mia Lehrer and Associates, TEN-Architectos, and artist Mary Miss) on 1,316 acres of a former U.S. Marine Corps base that contains four landfills of solid waste, paint residues, oily wastes, industrial solvents, and incinerator ash from military fabrication, production, operations, and maintenance activities.[3]

For each of those well-known park projects on vast industrial sites, there are dozens below the radar of the professional press. For instance, the recently closed 2,300-acre Lorton Reformatory—the former Washington, D.C., prison located south of the capital in Fairfax County, Virginia—is now subject to several planning initiatives, including the design of a large park that will incorporate the historic structures comprising the old prison community as well as a landfill of I-95 highway construction debris. Every metropolitan region in the country has a similar large parcel on its periphery that is slated for whatever redevelopment is deemed appropriate given its level of disturbance.[4]

Much of the writing about large parks on disturbed sites focuses on the processes of remediation necessary to cleanse them before human use can be considered safe. Although the eco-technologies and operational design strategies deployed in turning these wastelands into parks are fascinating and innovative, this particular focus fails to show what these large parks might mean to the communities that surround and use them. What does the large metropolitan park constructed on a site degraded by the processes of human consumption and industrial production mean? The urban institution known as the public park, once associated with landscapes affording urban dwellers respite from the world of work, consumption, and production, is now made on the detritus and the uncertain, perhaps toxic, byproducts of that realm. What is the social reception of uncertain parks that consist of circuit walks along metal boardwalks elevated above a toxic ground plane planted with heavy-metal-accumulating plants? Gathering places and memorials amid industrial ruins? Kite-flying mounds and extreme sports grounds situated atop tons of consumer waste, whose seepage is monitored? And large volumes of contaminated soil transported across the park for years as soil is cleansed, processed, stored, and then shaped into landform mounds for human use? What kind of citizens and society are associated with these large parks on disturbed sites? How do they compare with the citizens of a democratic society that were imagined to emerge from the experience of nineteenth-century American urban parks such as Central Park or Prospect Park?

FIG. 2: Color field of Acid Mine Drainage (AMD), acid-thriving algae, and "yellow boy," a rust-covered layer of metals that is the result of AMD, at the Hughes Borehole, Vintondale Park, Pennsylvania, coal mine.

Glancing Back at Large Park Precedents

How has the social conception of large public parks changed since the nineteenth and early twentieth centuries? In many ways, contemporary large parks perform the same roles: they afford city residents space to promenade and recreate in public, to immerse themselves in sun-drenched landscapes, to breathe clean air, to experience a vastness and character of space not generally found within narrow city streets and blocks. Historically, the large park's significance in the social construction of urbanity was predicated on the relationship between this public act of walking and the perceived healthfulness and size of its spaces.

Urban landscape was viewed through two lenses, a medical discourse and a social reform agenda. Herein lies a key similarity and distinction between the nineteenth-century and twenty-first-century large park, especially the large American park. If we look at the canonical parks of the nineteenth century, such as Frederick Law Olmsted and Calvert Vaux's Prospect Park, Central Park, and Franklin Park, we find large outdoor spaces where personal recreation took place amid nature. This resulted in what Olmsted called "unconscious or indirect recreation."[5]

But private reveries and public health were not the only effects Olmsted and Vaux desired from their public parks. Such experiences occurred in the presence of strangers—those from different walks of life. These park designers and their clients believed that spatial practices such as promenading, riding, and boating in the company of others engendered what Olmsted referred to as a sense of "communicativeness" or "commonplace civilization."[6] A democratic community emerged through the enactment of everyday recreational spatial practices in constructed rural scenery. That particular type of designed landscape evoked what these different people shared: their relationship to one another as equals, and their collective relationship to the vast continent of plentiful land that promised them health, productivity, opportunity, and belonging.

Spending time in nineteenth-century large parks meant opening up to the psychological and therapeutic effects of scenery, recognizing and empathizing with others, and reinforcing those bonds in relationship to the American landscape. This shared landscape was a visual and spatial register of natural beauty, abundant resources, productivity, regional pride, and national exceptionalism. Its presence in the city reinforced a sense of community and citizenship.

Granted, this ideological conception of the American large park is not the only one against which we can judge the new forms and meanings of contemporary parks. By the mid-twentieth century, most large parks were either large recreation machines full of programmed sports activities or large generic, pastoral landscapes. One park looked much like another, regardless of where it was built. This observation is widespread, as Paul Driver's beautiful meditation on parks in his "fictionalized autobiography," *Manchester Pieces,* demonstrates:

> I love to see an unfamiliar park—but parks almost by definition are primarily familiar—stretching out before me, lawny, rolling, crossed with avenues whose trees, let us say, are turning as autumn deepens by the minute, leaves already lending the avenues a patina of fading bronze at twilight.[7]

Large pastoral parks with ball fields and picnic shelters are more a form of amnesia, a practice of forgetting site histories, than indices of regional character and identity.

This ubiquitous, placeless recreational park or open-space park was the model for many of the first large parks built on disturbed sites. When I was a teenager living in Virginia Beach, Mount Trashmore, a large landfill within view of the toll booths on the new Virginia Beach expressway, was transformed into a recreation park with a lake and a sixty-foot-high mound. Hailed as "the first landfill park in the world" when it opened in 1973, there is little on site that speaks to its industrial history. Only those with astute environmental perception would wonder about the height of the hill, an incongruous figure in the flat, coastal region at the mouth of the Chesapeake Bay. That history—and the processes of settling, decomposition, remediation, and groundwater seepage that might accompany it—is hidden under a thin green veneer of grass and asphalt. Like the open-space parks of the mid-twentieth century, these early large parks on industrial sites were a form of forgetting and deception, designed with what Mira Engler calls the "camouflage approach."[8]

Glancing Sideways: Large Parks as Sites of Consumption as well as Production

This art of landscape camouflage masks the histories and processes of disturbed industrial sites and obliterates a connection that might render these parks more meaningful to the public. As remnants of twentieth-century industrial society, these sites have the capacity to tell stories about consumption as well as production. As the residue of collective consumption and mass production, slag heaps and below-ground chemical plumes are direct manifestations of the unacknowledged and largely unseen consequences of technological processes and industrial manufacturing. These landfills and waste landscapes are the epiphenomena of a society wherein individuals buy as much out of desire as need, and consume in response to marketing of identities as much as services provided. In Engler's words, "As the greatest earthwork monument of our times and a cogent symbol of our consumer culture...a dump can be held as a mirror to our culture."[9]

What kind of culture is this? What type of society creates such places? Surely not the same sort of "commonplace civilization" that found shared meaning in the rural scenery of its region or nation. Lizabeth Cohen describes twentieth-century America as a time when "mass consumption is the overriding cultural experience, much the way religion must have been in the seventeenth century, revolution in the eighteenth century, and industrialization a century ago."[10] Cohen persuasively argues that by the late twentieth century, "citizen and consumer had become interlocking identities for most Americans," and that this changed sense of citizenship was accompanied by new patterns of settlement.[11]

Thus American identity, at home and abroad, was tied more to the products, images, and landscapes associated with consumption and its display, and less to the scale and character of its immense, unbounded continental landscape. Those raised in vast, sprawling landscapes of highways, strip malls, and suburbs—spaces of mass consumption and display—are not going to find meaning in either the nineteenth-century rural scenery park or mid-twentieth-century open-space park.

One key to finding such meaning resides in Cohen's description of the "third wave consumerism" of the 1960s. During this decade, American "confidence that a prosperous mass consumption economy could foster democracy" was accompanied by the recognition that such an economy created pollution along with its products.[12] With mass production and consumption came the unintended destruction of ground, air, and water quality. Consumers began to demand protection from pollution at all scales, from unmarked additives in food products to pesticides in agricultural production to chemical spills in waterways to factory smokestack particulates in urban smog. They recognized the creative destruction of capitalism and acknowledged the interconnectedness of their roles as consumer-citizens, technology, industry, and the environment. Legislation and policy changed to protect their rights both to consume safely and to the environment necessary to sustain that consumer society.

Still, few designed landscapes exist from this period that give form to these debates, the values that undergird them, or the contradictions that give them such resonance. What might be the options for imagining large parks on uncertain, disturbed sites resonant with a consumer society that is aware of the connection between consumption, production, and pollution? Can the experience of such a large park reveal the gap between contradictory American conceptions of land as a source of reverie and a resource for industrial production? How is landscape as a cultural product and the park as a type redefined in this process?

Toward a Praxis of Designing Large Parks on Disturbed Sites: Working Assumptions

1. Large parks on disturbed sites should be recognized as landscapes of consumption as well as production. It is tempting for designers of large parks built on abandoned industrial sites to heroicize the buildings and machines that remain. Such strategies, however, privilege the histories of production over the histories of consumption that are also embedded in such sites.[13] This allows visitors to distance themselves from the histories of human, material, and chemical flows on and off the site, and to limit their own culpability in and responsibility for such histories. ("Malevolent industrialists polluted the air and water, not my ancestors and certainly not me.")

Similarly, design strategies that focus primarily on the ecological processes of remediating a toxic industrial site fail to account for the intermingling of the natural, social, and industrial processes that permeate such sites. Forests, earth, and rivers are processed into lumber, ore, and water that are the raw materials for industrial production. The results of the process are consumer goods and emissions into the ground and waterways. Technology doesn't simply transform nature into commodities; it cycles back new and often toxic byproducts into nature. Thinking about landscapes of consumption and production requires thinking of the circulation of need, desire, material, goods, energy, and waste across disciplinary categories such as nature and culture, ecology and technology, and even public and private.

We need design strategies that make visible the past connections between individual human behavior, collective identity, and these larger industrial and ecological processes. How might such approaches help us reimagine a society of consumers who are aware of their impact on our habitat, not simply out of self-interestedness but arising from a sense of extended, interconnected communities?[14] Perhaps they can give scale, size, and measure to our aggregated habits of individual consumption. Perhaps large parks can allow spatial practices that connect the personal act to the collective public disturbance, that allow park visitors to consider the difference between thinking "green" and acting "green," between one's values and behaviors.

This is one of the key problems with the environmental movement, and of extant strategies for remediating and reusing disturbed sites as public spaces.[15] Designers who do imagine the intertwined social and ecological processes of these large parks rarely view the social as more than a mass of singularities recreating together. They have not found ways to envision parks as agents in constructing a new kind of community based on relationships between humans and the land. Unlike statistics and reports, places can make palpable the consequences of processes that can seem abstract and disembodied, such as industrialization. Could designers of large parks make spatially legible the contradictions between broad social values such as environmentalism and individual habits such as consumption? If so, they might once again become places of social agency. They might be prompts for refashioning the definition of citizenship, from that of consumer-citizen or environmentalist-citizen to that of consumer as environmentalist-citizen.

2. Large parks on sites of consumption and production should be more than symbols of our consumer culture. They should be recognized as the invisible consequence of our needs and desires, and should be remade, in Engler's words, into "a place where our rejected and silenced cultural values are awakened, provoked, and interrogated."[16] This process might focus on the volume of material and energy that is moved and expended as a consequence of making, acquiring, using, and discarding the products of our desire and need. By giving form to such an inquiry, by expanding the understanding of a landfill mound from a sculptural mass to a temporal flow of matter and waste, we begin to overcome our visual conceptions of landscape that inhibit us from seeing the disturbed landscape for the place of circulation and exchange that it was and is. The landscape as scene, or view, separate from technology and industry, gives ways to what Barbara Adams calls a timescape. It is a way of seeing the environment and its invisible hazards that is rhythm-oriented and capable of discerning the synchronicities of industrial time, ecological time, and social time. Adams explains how a timescape can overcome binaries such as industry and ecology, culture and nature, society and individual that preclude seeing connections between processes and across systems. Her writing resonates with contemporary landscape architecture discourse and connects that design theory to environmental policy and ethical realms.[17] It prompts us to ask more from large parks on disturbed sites than remediation of contamination and provision of recreation.

3. A timescape conception of large parks leads to a recognition of uncertain sites—spaces where matter, flow, and waste know no boundaries—and to a different conception of consumer society. Large parks on former dumps, landfills, or factories that hold or made objects of our desires and needs are not separate from our places of home regardless of physical distance. Toxicity flows. It transgresses property lines, watersheds, and ecosystems. The edges between large parks and their adjacent properties are spaces for revealing the air, water, and energy flows across legal boundaries. The specific design forms that mark these liminal spaces between park and city, such as ha-ha walls, weirs, remediation plots, and monitoring wells, are also spaces of uncertainty. These moments should be spaces of witness. If contaminant containment is understood as uncertain, citizens might adopt a different conception of consumer society. They might understand the connections between the products they consume, the lifestyle they lead, and the polluted environment.

This speculation is not an ideal dream, but an existing reality. Lawrence Buell asserts that the anxieties of our post-industrial culture have resulted in a new "widely shared paradigm of cultural self-identification—toxicity." He describes this toxic discourse as "perceived threat of environmental hazard due to chemical modification by human agency," and notes that it permeates American life, news, and experience. He suggests that toxic discourse has already broken down the binaries of city and countryside, ecology and technology, and has resulted in a "disenchantment from the illusion of a green oasis."[18]

Buell's characterization of Americans' perceptions of their environment challenges what many landscape architects assume about their clients and popular culture. From his research, it appears that the "camouflage approach" to regenerating brownfield and Superfund sites—the terminology Americans typically use to describe degraded, uncertain industrial sites—is not only disingenuous in the way it hides the processes at work. More disturbing, this approach fails to reinforce the nascent community that exists between citizens and their disturbed environment. In Buell's words, "More and more it may become second nature to everyone's environmental imagination to visualize humanity in relationship to environment not as solitary escapes or consumers but as collectivities with no alternative but to cooperate."[19]

Buell finds that Cohen's consumer-citizens of the late 1960s and early 1970s have become, and raised children who are, consumer-environmentalist citizens. They deserve large parks on disturbed sites that make visible the uncertainties of those sites. This revelation, while connected to design theories of the 1990s, such as those discussed in the Landscape Journal special issue on "Eco-Revelatory Design," is a theoretical position with an equally strong aesthetic and ethical component.[20] Revelation is not only in the service of making ecological processes, or landscape design, visible. It is a tool for reinforcing the fear that is individually suspected—that the disturbed industrial landscape is a source of collective identity. Ulrich Beck has called this landscape our "shadow kingdom"; it is a processed, manufactured, industrialized, and contaminated landscape, one that is not out there but unbounded, everywhere, in here—in our water, soil, and air; in our suburbs, schoolyards, neighborhoods, and large parks.[21] We have shaped that toxic landscape through our individual behaviors, our collective patterns of consumption, and our collective nonsustainable settlement patterns.

4. Toxic discourse is an expression of a collectivity of consumer-citizens who perceive their environment through the lens of uncertainty and risk. Disturbed sites are the byproducts of economic policies that viewed nature as a resource and that accepted environmental degradation as the inevitable consequence of technological progress. The experience of designed landscapes on and in disturbed sites can render visible the consequences of the economic, political, and social decisions that led to those risks. After September 11, 2001, it is not profound to state that Americans live in a risk society.[22] This phrase initially referred to the environmental risks that are the result of political decisions about economic growth and industrial production. It was popularized in Ulrich Beck's writings in the 1980s and 1990s, especially Risk Society: Towards a New Modernity and Ecological Enlightenment: Essays on the Politics of the Risk Society. In these books and other writings, he extends the arguments made by environmentalists such as Rachel Carson and Ian McHarg in the 1960s, by illuminating the degree to which environmental policy is bound to the hazards assumed acceptable within industrial processes.

Beck implicates all of us in the acceptance of these risks when he writes, "A transformation of our way of life has taken place in the guise of undisclosed latent side effects."[23] We inherited this condition of toxic flows transgressing spatial and temporal boundaries because we wanted a certain quantity of things, a certain level of technological development and attendant cultural progress. We desired newer, more, faster, closer. And we knew there were risks, but we assumed they were small, containable.[24] Disturbed toxic sites are the result of specific industrial processes. But they are also the result of the way we wanted to live our lives, and our naive belief that our chances of exposure to danger were remote.

Disturbed sites are risk materialized, spatialized, and temporalized. As such, large parks on these uncertain sites have a significant role to play in a risk society, as they present so clearly the difficulty of containing the hazards of industrial processes. The flow of toxic chemicals out of containment pits, beyond clay caps and property lines, downstream and downwind can be read about and acknowledged. But it is known differently, and I would argue more deeply, through sight, scent, sound, and touch, the mind engaged with its body, the intellect connected to emotions. A consumer-environmentalist citizen who perceives the environment through a toxic discourse and who visits a large park undergoing remediation processes over a period of years will know her environment differently. And like Beck and Buell, I suspect, she will consider, and reconsider, the risks necessary to take to sustain her habits of consumption.

Toward a Praxis of Large Parks on Uncertain Sites: From Toxic Discourse and Individual Collectivities to Spatial Practices

Threats from civilization are bringing about a new kind of "shadow kingdom," comparable to the realm of the gods and demons in antiquity, which is hidden behind the visible world and threatens human life on this Earth. People no longer correspond with spirits residing in things, but find themselves exposed to "radiation," ingest "toxic levels," and are pursued in their very dreams by the anxiety of a "nuclear holocaust"…Dangerous, hostile substances lie concealed behind the harmless facades. Everything must be viewed with a double gaze, and can only be correctly understood and judged through this doubling. The world of the visible must be investigated, relativized and evaluated with respect to a second reality, only existent in thought and concealed in the world.[25]

Buell's chapter on "Toxic Discourse," from which the above passage is taken, begins with this quote by Beck describing a risk society. Their collective argument is apt, especially when merged with Cohen's, in characterizing the contradictions between Americans' individual values, their collective identities, and their actions. Still, I am compelled to suggest more. How can we construct a public realm that manifests and enacts relationships that are known—the shadow kingdom—but not acknowledged? That spatializes the connection between industry and community, consumption and waste/pollution, security and power, ecology and technology, humans and the biotic world? That is more than a free space for "anything goes," but avoids the trap of a designer overdetermining the meaning of a place?

This public realm for a risk society can be constituted around the interconnections between bodies, values, actions, industry, and the environment. It can be enacted and performed through intersecting spatial practices that combine program, duration, and site. Engler has come closest to describing this idea with her "integrative approach":

> It integrates the principles of ecology with the philosophy of art—scientific rigor with expressive metaphors. It is layered with information and meanings that express the dynamic balance of nature and culture.... It combines ecological, economic, and social agendas and merges the utilitarian with the experiential. It invites people to partake in highly sensory and transparent landscapes. Above all, it facilitates the flow of matter and the sustenance of vital waste landscapes.[26]

But her vision is not radical enough; it persists in using words such as "balance" that seem tied to pre-1970s theories of ecology and associated conceptions of human/nonhuman relationships; and it implies that meaning emerges out of artful, celebratory forms, instead of out of the experience, the everyday routines and exchanges, that might be enacted through spatial practices in a public place.[27] And because her subject is waste landscapes versus all disturbed sites, she does not address the value of acknowledging risk and attempting to register uncertainty spatially and physically.

What are the design alternatives to this anachronistic sense of natural harmony and balance? To seeking to assign forms with specific meaning? Again, Buell's and Beck's writings are instructive. Within Buell's conception of toxic discourse through which Americans understand their relationship to one another and to the environment, we might expect to emerge new social bonds, "collectivities with no alternative but to cooperate."[28] "Collectivities" is an interesting choice of words, when "community" might be expected. It implies that individuals might work together toward a goal, but they are not gathering together. They maintain their individuality but recognize their interdependence through their environmental values and actions, through their patterns of consumption and production. As Ulrich Beck and Elisabeth Beck-Gernsheim more guardedly state:

> [The public realm] no longer has anything to do with collective decisions. It is a question not of solidarity or obligation but of conflictual coexistence.... The decline of values which cultural pessimists are so fond of decrying is in fact opening up to the possibility of escape from the creed of "bigger, more,

better," in a period that is living beyond its means ecologically and economically. Whereas in the old value system, the self always had to be subordinated to patterns of collectivity, these new "we" orientations are creating something like a cooperative or altruistic individualism.[29]

What might a park program, or park experience, be that tapped into and perhaps reinforced this sense of individuality within a collectivity, conflictual coexistence, and cooperative or altruistic individualism? Surely more than the glimpses of generic fit, smiley-faced people skating, cycling, and recycling through emergent landscapes that permeate most manipulated two-dimensional digital images (created using Photoshop or other software) of large parks on disturbed sites. There should be as many approaches possible as types of disturbed sites and their surrounding neighborhoods.

To imagine the encounters and spaces that might be possible, new ways of drawing and modeling are necessary. Perhaps the best example of how drawing can help surface the temporal unfolding of program can be seen in the works of Anu Mathur and Dilip da Cunha. We need more designers who can deploy similar notational systems for imagining the complex temporal and spatial choreography of ecological process, cultural rituals, industrial production, and political agency. Developed by Mathur and da Cunha over the past decade through various design competitions and research projects, their notational diagrams for the Fresh Kills Park competition are a model that deserves broader application during the planning and design of disturbed sites. Such analytical diagrams might allow a designer, her consultants, and her clients to imagine new arrangements, juxtapositions, dislocations, and amplifications of events that in their incongruousness make evident the unspoken and unacknowledged risks and relationships we assume in exchange for our comforts. The 2006 ASLA student awards submissions included several extraordinary examples of notational systems apparently inspired by Mathur and da Cunha, as well as James Corner and Alan Berger. Brett Milligan's award-winning submission in the Communications category, "Navigating Bigness," was a rigorous, sophisticated depiction of the intersection of social, ecological, industrial, and economic flows in, out, and through an industrial landscape. It was especially interesting in the way it depicted multiple scales, from site to regional and global (FIGS. 3–9).[30]

This likely means the intermingling of activities that many would find unacceptable or problematic, such as access to ongoing phytoremediation plots, or housing on disturbed sites, or fast-food kiosks near test plots, or boating on water-treatment ponds. A few new "almost large" parks have demonstrated the power of such unlikely juxtapositions, such as San Francisco's Crissy Field, by Hargreaves Associates, where a fragile bird habitat is juxtaposed with a busy bay-front promenade that is used regularly by residents of adjacent neighborhoods. The publicness of such a landscape is reinforced by multiple visits to this type of park, suggesting the possibility of other programs,

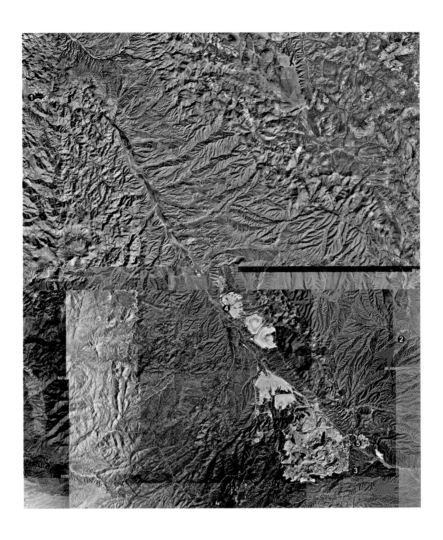

FIG. 3: This aerial of the Tyrone open-pit copper mine, entitled "Permutations and Extensions of Boundary," is the focus of Brett Milligan's University of New Mexico MLA thesis project, "Navigating Bigness: Redefining Corporate Landscape." His project, a hypothetical alternative corporate annual report, challenged the conception that industrial landscapes have fixed spatial boundaries by mapping the flows between this mining complex and its context, from the region to the globe. Figures 3–9 are a small sample of the diagrams and maps in "Navigating Bigness."

including daycare centers, schools, summer camps, eldercare facilities, and other activities that prompt repeated visits.

Yet the selection and choreography of the activities alone will not engender individual collectivity. These parks should afford encounters and experiences wherein the systemwide impact of one individual action or interaction after another is made palpable, and where one can imagine those actions and interactions transgressing systems, such as when an ecological process is absorbed into a technological process; or an observation and participation in a remediation technology occurs during an annual collective social event; or one's waste pile becomes a wildlife habitat. The experience of entering and leaving these parks, or walking and driving along their perimeter, offers one of the most challenging opportunities for designers. How can these spaces be released from their neutral role as buffers and be allowed to speak of the exchanges and overlaps? How can the park's boundary become a threshold where environmental consequences and risks are witnessed? The agency of a large park on a disturbed site depends on exploiting our interdependencies and on allowing us to enact our relationships as cooperative individuals, as consumer-environmentalist citizens of a risk society. We experience nature through our collective individual experiences and through a toxic discourse.

This goal may seem ambitious or naive, and difficult to realize given current regulations. As such, why should we bother when scholars such as Adams, Beck, Buell, and Cohen have already noted these changing sensibilities, identities, and affiliations in their writings? What does the large park do that one of their books, or an environmental impact statement, scientific journal, or experimental remediation plot can't do better? A large park on a disturbed site provides an immersive, aesthetic, collective experience in a vast landscape, one too large to grasp at a glance and so extensive that it implicates multiple systems and processes. This somatic, haptic, and yes, aesthetic experience transforms abstract knowledge into embodied knowledge. It has the capacity to move one to act in ways that reading might not.

Both Buell and Beck note the power of cultural products as well as scientific studies to alter environmental consciousness. Buell asserts that we should focus as much on "structures of thought, values, feelings, expression and persuasion" as we do on other forms of toxic discourse such as its chemical, medical, and legal aspects.[31] Again, Buell quotes Beck:

> The success of all environmental efforts finally hinges not on "some highly developed technology, or some arcane new science" but on a state of mind, on attitude, feelings, images, narratives, and "that only if nature is brought into people's everyday images, into the stories they tell, can its beauty and its suffering be seen and focused on."[32]

Here we begin to understand that the experience of nature inspires and provokes humans in ways that publications and facts cannot. The actions that

0 N 32.782625 W 108.075726

Santa Rita

Ninety-seven years of digging, blasting and hauling coalesce into a single moment.

The Bisbee/Lavender Pit

Bisbee was mined as an open pit for about 25 years, until 1974. Now the pit just sits, or rather slides. The entire southeastern slope has eroded from highway runoff, creating a smoothed, shifting plane. Deep rusty-orange puddles of chemical and copper-laden water disclose the remnant toxicity and transformation of regional water patterns.

0 N 31.463013 W 109.900156

Bisbee

4:30pm, February 17th, 2006

FIG. 4: "Navigating Bigness": The Santa Rita open-pit copper mine is part of a large corporate mining and manufacturing network that has displaced 4.5 billion tons of material to produce copper sheeting (top).

FIG. 5: "Navigating Bigness": A second open-pit mine, Bisbee/ Lavender, has been out of operation for thirty years, but the chemical and mechanical processes of erosion and pollution are ongoing (bottom).

result from such provocations are not predictable, of course. But they are different. And given the degree of denial and aversion that permeates environmental debates, new thoughts and actions are surely needed. In fact, probing, indeed extending our expectations for landscape beauty on disturbed sites, should go hand in hand with new spatial practices. For beauty in spaces of nature has the capacity to not only spur action but to shift our concerns from ourselves to the collective and our environment.

Elaine Scarry makes this case persuasively in *On Beauty and Being Just*, when she writes:

> At the moment we see something beautiful, we undergo a radical decentering. Beauty, according to [Simone] Weil, requires us "to give up our imaginary position as the center.... A transformation then takes place at the very roots of our sensibility, in our immediate reception of sense perceptions and psychological impressions." Weil speaks matter-of-factly, often without illustration, implicitly requiring readers to test the truth of her assertion against their own experience. Her account is always deeply somatic: what happens, happens to our bodies. When we come upon beautiful things...they act like small tears in the surface of the world that pull us through to some vaster space...we find we are standing in a different relationship to the world than we were a moment before. It is not that we cease to stand at the center of the world, for we never stood there. It is that we cease to stand even at the center of our own world. We willingly cede our ground to the thing that stands before us.[33]

This final point leading toward a praxis of designing large parks on uncertain sites implicates somatic, aesthetic landscape experiences—not pleasing, generic scenery—as a means to construct relationships and establish obligations between an individual and the world. We are challenged to imagine what kind of beauty might be appropriate, how it might mix pleasure and disturbance, and what sort of landscapes this implies. Two related examples might prompt additional speculation. Buell presents the first by quoting from Cheryl Foster's "Aesthetic Disillusionment, Ethics, and Art":

> If I am witnessing a spectacularly-coloured sunset from my kitchen window and am taking great pleasure in its beauty, how shall I respond when a friend of mine drops in and informs me that the reason for all the colour is the proliferation of sulphur dioxide in the air? Suppose that the friend also tells me that the sulphur, the result of a factory operating up river, is a pollutant, one with grave consequences for the creatures in the marsh downstream?[34]

This evocation of the paradox presented by this sulfurous sunset reminds me of my own reactions to the industrial landscape photographs of Emmet Gowin and Edward Burtynsky. The former's soft, silvery surfaces and dark, shadowy forms are gently mesmerizing, until their subject emerges out of the surface: mines, troop trenches, and nuclear test craters. The latter's saturated close-ups of smooth, striated marble outcrops and orange-colored liquids are also beautiful. But they are paired with distant panoramas of the same landscapes revealing the extent of environmental destruction caused by the industrial processes of mountain excavation. Do I change my mind about industrial processes because they cause beauty? Well, yes. I admit a certain guilt in saying yes. But that beauty also causes me to care about those mountains and valleys in the panoramic photographs in ways I did not before, and to a degree that reading about them would not.[35]

Axioms for Designers of Large Parks on Uncertain Sites: From Remediating Contaminants to Reimagining Community

Cohen's, Buell's, and Beck's scholarship suggests that the audience already exists for new experiences of the designed landscape in the form of large parks. It is an audience capable of connecting sites that are byproducts of sprawl, consumption, and production with historical agency and contemporary life. It is an audience whose very bodies and collective actions transgress the boundaries between health and toxicity, ecology and technology, past and present. Yet the unquantifiable uncertainty of risk attendant to playing on and moving through these sites remains abstract to this audience. Thus we can make the case for the agency of parks versus other forms of knowing.

Large parks provide opportunities to remediate large landscapes. But they offer even more—the opportunity to change environmental attitudes, construct new constellations of social collectivities, and prompt actions. The former industrial sites upon which these large parks are built are agents of toxicity, from the residue of production and consumption. If one accepts Cohen's thesis that since the 1950s in the United States progress has been equated with mass consumption and citizenship with consumption of those goods, these sites carry with them powerful associations. Factories, power plants, and especially landfills are not just symbols of obsolete technology. They are reminders of that which was shared in late-twentieth-century American culture: affluence, acquisition, planned obsolescence, and waste. The recycling of these sites into new urban and suburban institutions, new public parks, can be interpreted along a continuum—as acts of redemption and validation of those values, or as monuments to our collective excess and hubris. If our collective identity has been shaped by our shared patterns of consumption, and those patterns have led to unintended environmental degradation, the reuse of degraded sites as parks might resurrect the parks' agency as a vehicle for engendering new connections between private actions and public values, between individuals and the world. Large parks are places where consumer-citizens are allowed the opportunity to recognize the gaps between their alleged environmental values and their patterns of consumption;

① N 32.865228 W108.579202

Tyrone

Synergy

The sheer size of Phelps Dodge's mining operations creates an economic environment that fosters and necessitates the growth of other corporations. This is a specific quality of extraction industries that generate entire towns and economies based on the labor and energy requirements of their operations.

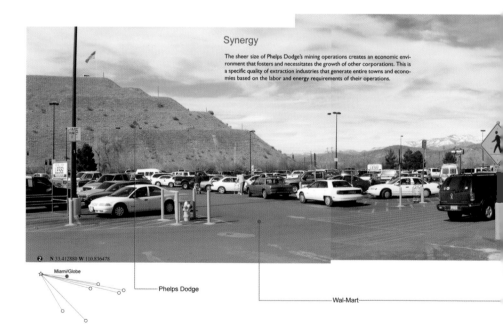

② N 33.412880 W 110.836478

Miami/Globe

Phelps Dodge

Wal-Mart

Bill Evans Lake is over twelve miles northwest of the Tyrone mine, yet it was specifically built to supply the mine with water drawn from the Gila River. The water is pumped several hundred feet up to the artificial lake via a pumping station next to the Gila, with a storage capacity of 2100 acre feet of water. The lake gets its name from the lawyer who enabled Phelps Dodge to build the infrastructure in 1969 by selling it as a public amenity. The steep semi-circular earthen dam at the southern edge of the lake ensures the visual impossibility of it being a "natural" lake. The water is stocked with fish and the Saturday evening that I was there people were fishing and hanging out by their RVs. Signs warn against swimming in the lake as someone drowned here in 2003. I wondered if it had anything to do with the massive pumps and pipes that emit the water somewhere out near the middle. A fascinating, largely invisible, engineered phenomena. Bill Evans Lake doesn't appear anywhere in Phelps Dodge's recent annual reports.

McDonald's   USA Today

FIG. 6: "Navigating Bigness":
Bill Evans Lake is a residual
recreational space and a
disturbed site. In this abandoned
open-pit mine, boating is allowed
but swimming is not, as danger
lurks below the surface in the
form of an industrial water-
pumping mechanism (top).

FIG. 7: "Navigating Bigness":
Large industrial sites such as
these open-pit mines require
adjacent housing for workers
that are in turn part of corporate
ecologies (bottom).

The in-between landscape

The distilled product is displaced from the corporate site of its origin to be purchased and used some where else. The in-limbo process of getting to the buyer involves an elaborately orchestrated syste of transportation flows in which the commodity glides along public highways to its destination.

❶ N 33.448819 W 112.073442

Phoenix

Surprisingly, **there is no landscape facade.** The only landscape here is a small curb and gutter parking lot behind the tower. The front of the structure leaves no room for ornamental landscape as it edges right up against the street. The **corporate facade occurs in the slick vertical planes of a tower** that fully occupies the entire block. One doesn't walk around the building, but rather through its edges. Phelps Dodge **references its productive process through the decorative display of the commodity it produces.** Copper (and facsimiles of copper) in painted plastic, guilded edges and surfaces of the tower. The glass facade reflects outward, rather than inviting one in, creating a fortress that is **visually impenetrable.** The effect is a near perfect built analogy for the secretive strategic positioning of Phelps Dodge's upper and senior administration.

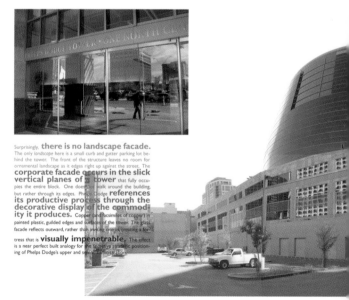

Accepted programmatics of the urban condition:

The everyday clean, crisp assemblage of industrial materials. The invisibility of accumulated landscape pastiche. Glass, steel, stone, concrete, copper...
The enabling systems of material hypermobility are so ingrained that they go unseen.

FIG. 8: "Navigating Bigness": Disturbed industrial sites are connected to consumers and communities through a vast transportation infrastructure network (top).

FIG. 9: "Navigating Bigness": The corporate headquarters, the symbol of many manufacturing enterprises, is a mute image yielding no clues about the vast network of material and energy flows required to produce its thin, copper veneer (bottom).

within a large park on a disturbed site, a generation of consumer-environmentalist citizens might find physical manifestations of their contradictory beliefs, impulses, and actions.

Large parks are unique cultural products and environmental experiences that can make visible the unspoken and invisible relationships between abstract systems. This requires conceiving of a large park as more than a big thing, or particular place, or emergent system, but as an aesthetic experience of unspoken and invisible relationships that can overcome the "blindness of everyday life" and the self-centeredness of consumer society.[36] It urges us to imagine the large park as the multitudinous field of individual spatial practices whose intersections and overlaps improve one's attentiveness and offer one a sense of collective activity and "conflictual coexistence." In making these processes and the threats or risks associated with them visible or palpable, individuals might acknowledge them and be motivated to act.[37]

To be effective as an agent of changing collective identity and environmental values, large park design must employ multiple, temporal strategies. This is not a new concern in landscape architecture. It is intrinsic to the landscape medium. It was essential to Lawrence Halprin's writing and designs for the last forty years. Many design critics and theorists, including me, have commented on the shift from spatial to temporal preoccupations in landscape architecture theory and practice since the late 1980s. More recently, most premiated entries in large park competitions, from Landschaftspark Duisburg-Nord to Fresh Kills to Downsview Park, have employed design strategies that exploited the temporal qualities of the landscape as a dynamic, performative, open-ended process medium.

What is different about this call for the temporal is that is carries with it an ethical agenda that is especially germane on industrial sites with histories of contamination that are undergoing biological remediation. Adams suggests that modern society was unable to conceptualize the temporal dimension of industrialization's impact on the environment. Industrial time—the time of calendars, clocks, and machines—dominated, and was considered separate from the dynamic processes of nature. Adams brings this critical stance one step closer to the landscape architecture design of uncertain sites by suggesting that the key conceptual tool for overcoming our current environmental problems is temporality, not spatiality.[38] Her introduction to *Timescapes of Modernity: The Environment and Invisible Hazards* might easily be read as an explanation of why post-industrial disturbed sites are so polluted:

> Nature, the environment and sustainability, however, are not merely matters of space but fundamentally temporal realms, processes and concepts. Their temporality, furthermore, is far from simple and singular. It is multi-dimensional, a multiplex aspect of earthly existence. Without a deep knowledge of this temporal complexity, I suggest, environmental action and policy is bound to run aground, unable to lift itself from the spatial dead-end of its own

making.... I focus on the conflicts that arise within the industrial models of life from a) the complexity and interpenetration of rhythms: cosmic, natural and cultural; b) the imposition of industrial time on the rhythmicity and pace of ecosystems; and c) the prevailing emphasis on visible materiality and quantity at the expense of what is hidden from view and latent.[39]

She continues, more directly addressing the topic of landscape:

Where other scapes such as landscapes, cityscapes, and seascapes mark the spatial features of past and present activities and interactions of organisms and matter, timescapes emphasise their rhythmicities, their timings and tempos, their changes and contiguities.[40]

Although I suspect that the last thing landscape architectural design criticism needs is another form of "scape," Adams's concept overtly avoids the spatial and visual bias of "landscape." Timescape, more explicitly than site, intertwines ecological and industrial processes that unfold over time. I refer to it here not so much to add another "scape" to design vocabulary but to underscore that spatial practices called for earlier in this essay assume everyday routines, gestures, bodily movement, rhythms, and duration. Their temporality is multiple in that these spatial practices traverse matter and surfaces that are altered by dynamic, hybrid ecological and industrial processes.[41] This enriched conception of temporality acts as a conceptual bridge between so many seemingly unrelated categories—body and site, ecology and industry, consumption and production. It is a key to establishing a praxis—a critical, theoretically based practice—of large park design on uncertain sites.

Strange Beauty, Landscapes of Risk, and Cultural Agency

The challenges of designing large parks on uncertain, disturbed sites have been taken on as a technical challenge by numerous landscape architects. The necessity of relying on processes that take place over long periods of time have led others to revel in the operational, the performative, the temporal aspect of these large, complex sites. I am convinced that these large parks can be so much more. They can contribute to and challenge civic life, toxic discourse, and our consumer republic. They can assist in reframing environmental problems as social, economic, and political issues that implicate industrial processes.

In doing so, large parks function similarly to other types of cultural products, from painting and films to eco-criticism and science fiction. They register our environmental doubts, insecurities, and dreams, at the same time expanding our understanding of the environmental consequences of our consumer republic and risk society. But large parks contribute differently than other cultural products because they so thoroughly alter the very substance that is their subject, the disturbed landscape. Large parks affect us differently in that we experience the particular configuration of their physical forms through our bodies and through spatial practices dispersed across vast ecological fields or landscape matrices. Therein lies the unique role and promise of large parks.

A peculiar landscape beauty can be found on contaminated sites, such as Landschaftspark Duisburg-Nord's yellow moss atop slag heaps, or its forest of birch trees and its thickets of willow and butterfly bush colonizing contaminated railroad right-of-ways. It can also be amplified, condensed, or created on remediated contaminated sites, such as when Peter Latz designed children's playgrounds and gardens within steel-sided bunkers that previously stored industrial materials, such as coke and limestone.[42] A peculiar landscape beauty is condensed in the startling color progression of the acid mine drainage-treatment basins at Vintondale Park, Pennsylvania. There, the coal mine drainage turns from "acidic orange to pea green to alkaline blue," as it is progressively cleaned in a series of discrete but interconnected basins (SEE FIG. 2).[43]

Elaine Scarry's *On Beauty and Being Just* prompts designers of disturbed sites to recognize that simply cleaning a site is not enough. Creating beauty—out of the strange, particular character found on contaminated industrial sites—is the first step in the process of environmental recentering. The challenge for designers of disturbed sites is Scarry's claim that the beauty that recenters, destabilizes, and moves us to care about "the other"—the beauty that has agency—is not generic or familiar. It is always particular. How could one argue for the "camouflage approach" to disturbed sites after knowing this?

Landscapes of disturbance will always be sites of uncertainty and risk; places of anxiety and discomfort. The experience of beauty that is found and created there will echo with the pervasive unease and insecurity citizens have about the shadow kingdom that we share, our contaminated American landscape. This new public realm of disturbed and disturbing sites is in need of witnesses who recreate, live, and work in and around their vast spaces. Witnesses who encounter landscapes of disturbance, doubt, uncertainty, and beauty in their everyday experiences of a large park might be bewildered, moved to wonder, and recentered. They might even ponder the private as well as public dimensions of citizenship. What might happen if that experience of beauty within risk caused a collectivity of individuals to act differently in their everyday lives? We might truly know what the cultural agency of landscape could be.

NOTES

1. DIRT Studio's Julie Bargmann has titled several of her lectures about her design practice, which focuses on contaminated or toxic sites, "Toxic Beauty." See also Mira Engler, *Designing America's Waste Landscapes* (Baltimore: Johns Hopkins University Press, 2004), and Niall Kirkwood, *Manufactured Sites*: *Rethinking the Post-Industrial Landscape* (New York: Spon Press, 2001).

2. "A disturbance is any relatively discrete event in time that disrupts ecosystem, community, or population structure and changes resources, substrate viability, or the physical environment." See Steward T. A. Pickett and P. S. White, eds., *The Ecology of Natural Disturbance and Patch Dynamics* (Orlando: Academic Press, 1985), 7; and Steward T. A. Pickett, Victor T. Parker, and Peggy L. Fiedler, "The New Paradigm in Ecology: Implications for Conservation Biology above the Species Level," in Peggy L. Fiedler and Subodh K. Jain, eds., *Conservation Biology* (New York: Chapman and Hall, 1992), 66–88.

3. El Toro Marine Corps air station website: http://www.globalsecurity.org/military/facility/el-toro.htm (accessed Aug. 31, 2006). Gordon Smith, "A Park Like No Other," *The San Diego Union-Tribune* (May 7, 2006): http://www.signonsandiego.com/uniontrib/20060507/news_mz1h07park.html (accessed Aug. 31, 2006).

4. A final indication of the timeliness of developing these post-industrial sites is the increase in the number of university programs in architecture and landscape architecture offering studios, technical courses, and seminars on this topic. I was on the 2006 ASLA student awards jury and reviewed numerous exemplary design, research, and communications submissions dealing with ways to describe, analyze, and design within disturbed sites.

5. Charles Beveridge and Paul Rocheleau, *Frederick Law Olmsted: Designing the American Landscape* (New York: Rizzoli, 1995), 35.

6. Beveridge and Rocheleau, *Frederick Law Olmsted*, 49–50; and Adam Gopnik, "Olmsted's Trip," *The New Yorker*, March 31, 1997, 96–104.

7. Paul Driver, *Manchester Pieces* (London: Picador Press, 1996), 156.

8. Engler, *Designing America's Waste Landscapes*, 37; and Mira Engler, "Waste Landscape: Permissible Metaphors in Landscape Architecture," *Landscape Journal* 14, no. 1 (1995): 10–25.

9. Engler, *Designing America's Waste Landscapes*, 123.

10. Lizabeth Cohen, "The Mass in Mass Consumption," *Reviews in American History* 18 (1990): 548–55.

11. Lizabeth Cohen, *A Consumer's Republic: The Politics of Mass Consumption in Postwar America* (New York: Knopf, 2003), 408–9.

12. Ibid., 347, 404.

13. The history of labor and local community histories are also implicated in these disturbed sites. While not the focus of this essay, they are also histories that are often overlooked in contemporary design practices. A notable exception has been the University of Virginia landscape architecture design studio collaborations of Julie Bargmann and architectural historian Daniel Bluestone.

14. Cohen, *A Consumer's Republic*, 408–9.

15. See Nicholas Nash and Alan Lewis, "Overcoming Obstacles to Ecological Citizenship: The Dominant Social Paradigm and Local Environmentalism," in Andrew Dobson and Derek Bell, eds., *Environmental Citizenship* (Cambridge, MA: MIT Press, 2006), 153–84, especially "Cultural Obstacles to Ecological Citizenship," 155–58.

16. Engler, *Designing America's Waste Landscapes*, 123.

17. Barbara Adams, *Timescapes of Modernity: The Environment and Invisible Hazards* (New York: Routledge, 1998).

18. Lawrence Buell, *Writing for an Endangered World* (Cambridge, MA: Belknap Press of Harvard University, 2001), 53, 31, 34.

19. Ibid., 53.

20. *Landscape Journal* 17, no. 2 (1998). "Eco-Revelatory Design: Nature Constructed/Nature Revealed" was guest edited by Brenda Brown. It included a few projects on disturbed sites, such as Kristina Hill's "Ring Parks as Inverted Dikes" and Julie Bargmann and Stacey Levy's "Testing the Waters," but its primary focus was "design that reveals and interprets ecological phenomena, processes and relationships—what we are calling eco-revelatory design."

21. Ulrich Beck, *Risk Society: Towards a New Modernity* (London: Sage Publications, 1992), quoted by Buell, *Writing for an Endangered World*, 30. Beck describes risk society as "an epoch in which the dark sides of progress increasingly come to dominate social debate. What no one saw and no one wanted—self-endangerment and the devastation of nature—is becoming the motive force of history. Here we are not concerned with analyzing hazards but with proving that new opportunities for arranging society arise under the pressure of the industrial threat that humanity will annihilate itself and the breakup of social classes and social contrasts that it causes.... Unlike the risks of an early industrial society, contemporary nuclear, chemical, ecological and biological threats are (1) not limitable, either socially or temporally...." Beck, *Ecological Enlightenment: Essays on the Politics of the Risk Society* (Amherst, NY: Humanity Books, 1995), 2. This definition is tied to the consumption, production, and disposal of goods as well as to the uncertain risk of toxicity: "Industrial society was defined by the distribution of goods, says Ulrich Beck, but our newest stage of history, is defined by distributions of risk, hazard, and danger. The best way to characterize this new stage is a 'reflexive modernization,' where reflexive means that 'the advancement and the dissolution of industrial society coincide.' Beck theorizes this new modernization by focusing on the danger, the side effects of production, and the toxic assaults that nowadays seem so prevalent." Lee Clark, review of Beck's *Risk Society*, in *Social Forces* 73, no. 1 (Sept. 1994): 328.

22. The Museum of Modern Art in New York City presented "SAFE: Design Takes on Risk" (16 October 2004–2 January 2005), an exhibition of "300 contemporary products and prototypes designed to protect body and mind from dangerous or stressful circumstances, respond to emergencies, ensure clarity of information, and provide a sense of comfort and security." The familiarity of so many of its objects, from carry-out coffee cup holders to baby carseats to childproof medicine bottles, also exposed how much we accept risk in our lives. See http://www.moma.org/exhibitions/2005/safe.html (accessed Sept. 1, 2006).

23. Beck, *Ecological Enlightenment*, 66.

24. William Leiss's online book review of *Risk Society* in the *Canadian Journal of Sociology* summarizes Beck's argument cogently: "our fate is bound up with risks that are deliberately undertaken—for the sake of benefits conceived in advance—by means of technology over nature." http://www.ualberta.ca/~cjscopy/articles/leiss.html (accessed Sept. 1, 2006).

25. Buell, *Writing for an Endangered World*, 30.

26. Engler, *Designing America's Waste Landscapes*, 40.

27. For more on the concept of spatial practices, see Michel de Certeau, *The Practice of Everyday Life* (Berkeley: University of California Press, 1984); and Henri Lefebvre, *The Production of Space* (Cambridge, MA.: Blackwell Press, 1992).

28. Buell, *Writing for an Endangered World*, 53.

29. Ulrich Beck and Elisabeth Beck-Gernsheim, *Individualization: Institutionalized Individualism and its Social and Political Consequences* (London: Sage Publications, 2002), 26, 28.

30. The best source on the Fresh Kills competition entry by Anuradha Mathur and Dilip da Cunha (in collaboration with Tom Leader Studio) is *Praxis: Journal of Writing and Building* 4 (2002):40–47. Other sources for their notational systems and site mappings include Anuradha Mathur and Dilip da Cunha, *Mississippi Floods: Designing a Shifting Landscape* (New Haven, CT: Yale University Press, 2001) and *Deccan Traverses: The Making of Bangalore's Terrain* (Delhi: Eastern Book Corporation, 2006). Brett Milligan, now living in Portland, Oregon, completed "Navigating Bigness" while a graduate student at the University of New Mexico.

31. Buell, *Writing for an Endangered World*, 31.

32. Ibid., 1; Beck, *Ecological Enlightenment*, 14

33. Elaine Scarry, *On Beauty and Being Just* (Princeton, NJ: Princeton University Press, 1999), 111–12.

34. Buell, *Writing for an Endangered World*, 212. Cheryl Foster, "Aesthetic Disillusionment, Ethics, and Art," *Environmental Values* 1 (1992): 205–15.

35. Lori Pauli, *Manufactured Landscapes: The Photographs of Edward Burtynsky* (Ottawa: National Gallery of Canada, 2005). Jock Reynolds, *Emmet Gowin: Changing the Earth* (New Haven, CT: Yale University Press, 2002).

36. Beck, *Ecological Enlightenment*, 13, 15. Scarry, *On Beauty and Being Just*, 109.

37. Beck, *Ecological Enlightenment*, 11.

38. Adams's argument resonates with those made by Beck. "The environmental problem is by no means a problem of our environs. It is a crisis of industrial society itself, reaching deeply into the foundations of institutions..." In other words, reconciling industrialization, pollution, and environmentalism requires a reconceptualization of landscape as timescape in addition to policies and programs. Beck, *Ecological Enlightenment*, 127.

39. Adams, *Timescapes of Modernity*, 9.

40. Ibid., 11.

41. Tim Ingold's "The Temporality of the Landscape," *World Archaeology* 25, no. 2 (October 1993): 152–74 is another example of recent scholarship that works out of the visual landscape's theoretical dead end. It recalls much of Anne W. Spirn's writings about time, rhythm, and landscape beginning in the mid-1980s, albeit from a different disciplinary bias.

42. Brenda Brown, "Reconstructing the Ruhrgebiet," *Landscape Architecture* 91, no. 4 (April 2001): 66–75, 92–96.

43. Bargmann and Levy, "Testing the Waters," 40.

FIG. 1: Boating on Fresh Kills wetland in the late nineteenth century.
FIG. 2: Fresh Kills Landfill, aerial view showing scale.

MATRIX LANDSCAPE:

CONSTRUCTION OF IDENTITY IN THE LARGE PARK

Linda Pollak

*The future belongs to the impure...to those who are ready to take
in a bit of the other, as well as being what they themselves are.*
—Stuart Hall

New York City's last landfill closed in March 2001 after more than fifty years
of operation (FIG. 2). When it reopened briefly six months later to accept 1.4 million tons of debris from the destruction of the World Trade Center, the international design competition for its redevelopment was already underway. Six
interdisciplinary teams had visited the 2,200-acre site during the first week of
September to initiate conceptual design and planning proposals for the park.[1]

Field Operations, the winner of the competition, is collaborating with
the New York City Department of City Planning and other city agencies to
create a process that will guide the site's development over the next thirty
years. In April 2006, Mayor Michael Bloomberg and City Planning Director
Amanda Burden announced the release of the Draft Master Plan for Fresh Kills
Park. According to the official project website, "the Parkland at Fresh Kills
will be one of the most ambitious public works projects in the world, combining state-of-the-art ecological restoration techniques with extraordinary settings for recreation, public art, and facilities for many sports and programs
that are unusual in the city."[2]

Taking the size of the Fresh Kills landscape into consideration, three
aspects of the site's identity contribute to its complexity: its historical use as
a landfill, its urban position, and its wetland ecologies.[3] What is at stake is the
construction of the identity of the park in ways that acknowledge and enable
difference. The folding of a multiplicity of social and natural concerns—from
multiple ecologies to multiple constituencies—into a landscape is a way of
affirming its heterogeneity. This heterogeneity, in support of resilient social
and natural landscapes, suggests a construction of identity that does not rely
on a single unified vision for a park. The argument is not against unity but
rather about the capacity to unify disparate and multidimensional areas of a
site and program without collapsing them into a monolithic identity.

Fresh Kills landscape, by being too large and complex to be fully com-
prehensible, operates as a catalyst for approaches that challenge established
aesthetic traditions of park design.[4] This challenge occurs particularly with
regard to the dominant paradigm of the pastoral park, as it developed in
North America in the nineteenth century, with a debt to the late-eighteenth-

surface water
fresh kills
glacial cretaceous soil
sandstone / shale
serpentine / schist

FIG. 3: Fresh Kills Landfill in a regional context, aerial view and section.

century English landscape garden. With its smooth surfaces, minimal program, and harmonious naturalistic environment, presented in contrast to the city as an antidote to the stress and artificiality of urban life, the pastoral park is an inadequate paradigm through which to address the challenges of the Fresh Kills project.

The design of Fresh Kills, in the competition entries and in the development of the winning "Lifescape" project, is the filter for this discussion. The issues, however, are relevant not only to the consideration of other large parks but to any project for which the illusion of a stable "whole" precludes the development and articulation of dynamic and heterogeneous aspects of identity.

The six teams in the competition took different approaches toward developing an aesthetic-environmental program for the site. The most successful of the competition entries outlined frameworks through which to address the differentiation of the site's conditions as processes to engage in their specificity, rather than problems to be solved at an abstract level. Such frameworks enable the acceptance of shifting identities and lack of containment in time and space, and the acknowledgment that a landscape cannot be fully controlled or even defined.

The mythical status of Fresh Kills, the world's largest landfill, as one of two human-made structures visible to the naked eye from outer space (the other being the Great Wall of China), suggests the unfathomable scale of the site, and in particular of the four mounds of accumulated garbage that rise out of the surrounding wetlands to heights ranging from 90 to 225 feet. A pair of numbers identifies each of the mounds, reflecting the merging over time of eight initial mounds. Mound 1/9, at 50 acres the largest as well as the most recent, is the location of debris from the World Trade Center, now the site of a planned memorial.[5]

The size of the site means that it has a significant impact on its context in ecological terms, especially because its wetlands support a great diversity of living systems at multiple scales. Its location on the Hudson River estuary heightens the role of its size, for example, in serving migratory routes. The site is bordered on the west by the narrow Arthur Kill, separating New York and New Jersey. To the east is the Staten Island Greenbelt, 2,800 acres of connected natural areas (FIGS. 3, 4).

The site is also intertwined with a substantial urban territory of varying densities and types of settlement, industry, and development. It is bordered by oil tanks of the industrial New Jersey waterfront to the west, residential neighborhoods including Travis and Arden Heights to the north and south, and the Staten Island Mall to the east. In terms of transportation infrastructure, the site is bisected by the West Shore Expressway, one of the major north-south traffic arteries of the island. Along its perimeter are the Arthur Kill Road to the south, Richmond Avenue to the east, Travis Avenue to the north, and Victory Boulevard to the northwest.

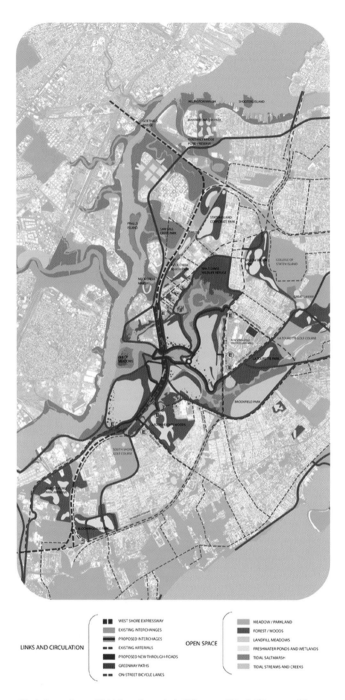

LINKS AND CIRCULATION

- ■■ WEST SHORE EXPRESSWAY
- EXISTING INTERCHANGES
- PROPOSED INTERCHANGES
- ■ ■ EXISTING ARTERIALS
- PROPOSED NEW THROUGH-ROADS
- GREENWAY PATHS
- ■ ■ ON-STREET BICYCLE LANES

OPEN SPACE

- MEADOW / PARKLAND
- FOREST / WOODS
- LANDFILL MEADOWS
- FRESHWATER PONDS AND WETLANDS
- TIDAL SALTMARSH
- TIDAL STREAMS AND CREEKS

FIG. 4: James Corner/Field Operations et al., "Lifescape," Fresh Kills competition proposal, diagram showing ecological relationships.

Notwithstanding the site's distance (and its neighbors' self-proclaimed difference) from the four more densely urbanized boroughs of New York City, Fresh Kills is inescapably urban, whether such urbanity is latent or acknowledged. Its development at this time is linked to the decentralization of investment by the city, inseparable from ongoing processes of urbanization of the region. Lifescape will be an urban park, subject to intensive recreational, educational, and other uses by people from throughout New York City and beyond, as well as those residents of adjacent neighborhoods who endured its previous identity as a landfill.

A visceral aspect of Fresh Kills' urbanity has to do with its historical material identity as a landfill for all of New York City: it was constructed from the detritus of the city's five boroughs, displaced to the site day after day (thousands of tons a day), truckload by truckload, barge by barge, for half a century. Some part of something now embedded in Fresh Kills has touched every inch of New York City, producing an urban social identity intensified by the integration of the World Trade Center debris.

Putting garbage out of sight, far away from the city center, made it possible to ignore it. To the extent that the Fresh Kills landscape is a consequence of our own material desires and consumption, its location away from the city hub reflects a desire to forget about our own waste products, to look in a different direction rather than risk being identified with them, to have them go away. While a visit to the site ten years ago would have made it clear that there is no "away," the subsequent burial of the landfill's entire contents seems to make it possible now to pretend otherwise, in several of the competition entries, notwithstanding the prevalence of post-closure infrastructure.

The fact that Fresh Kills is a brownfield site—as are the majority of large sites available today to be transformed into parks—increases its complexity in terms of the variable state of disturbance of its ecological systems, as well as the amount of existing and planned engineering.[6] Despite the placidity of its visible surfaces, the site is severely degraded and toxic. The mounds are characterized by leachate, off-gassing, and differential settlement—covered with a thin lid of plastic and soil.

As recorded in a video about the landfill, Mierle Laderman Ukeles, the Percent for Art artist of Fresh Kills who has been involved with the site for thirty years, interviews an engineer, who describes the process of vacuuming methane gas out of wells sunk into the landfill. This process must be carefully regulated because too much suction will bring in oxygen, upsetting the anaerobic process of methane production and producing underground fires. The engineer's narrative reveals that the landfill is not an inert mass, but a living and breathing, sometimes frightening landscape, to be respected and cared for (FIG. 5).[7]

Brownfield sites, with their histories of inaccessibility as industrial areas, landfills, unofficial dumping grounds, or military bases, present design challenges that go beyond decontamination. Each transformation

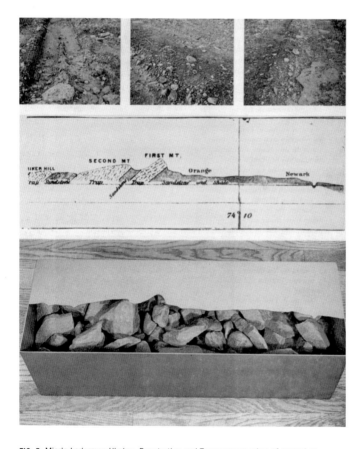

FIG. 5: Mierle Laderman Ukeles, *Penetration and Transparency*, view of ground at
Fresh Kills with pipes embedded (top).
FIG. 6: Robert Smithson, *Six Stops on a Section*, 1968. Art © Estate of Robert
Smithson/Licensed by VAGA, New York, NY (bottom).

of a brownfield to a park constitutes an inversion of its perceived value. Particularly at the edges of a site, this inversion suggests the importance of recognizing a site's historical identity, in the context of a need for significant urban design thinking, to help adjacent communities reengage with a site that has been isolated, inaccessible, or derelict, and has therefore existed as a barrier. Contrary to the desires of the site's neighbors, the project's acknowledgment of the site's history of abuse can help to make new sense of it as part of the production of a new social-natural identity.

The pastoral park's repression of disorder and complexity makes it an inadequate paradigm for dealing with brownfield sites such as Fresh Kills. Yet its attraction is undeniable, given the strong desires of neighbors and others to forget what has happened at such sites. The allure of a pastoral landscape approach lies in its potential to camouflage undesirable "unnatural" conditions. The power to bring abused land "back to life" is often intertwined with a desire for redemption, in a scenario in which a park becomes a symbol of healing. The idea of renewal of such a site is tied to the long tradition of using landscape to renaturalize land that bears the scars of exploitation, in the service of a more general urge for purification of not only that landscape but also of the individuals or groups associated with it, and therefore somehow represented by it. This redemptive urge emerges in the competition materials by omission, in rhetoric that sticks to "active recreational needs" and the "delicate ecology of the site."[8] The lack of representation by most of the competition entries of what happens beneath the mounds' grassy surfaces goes along with a tendency to define Fresh Kills primarily as a natural preserve, amplifying wetland phenomena and downplaying visibly constructed elements.

The idea that a site such as Fresh Kills could be returned to a natural condition perpetuates the myth that nature is separate from people, culture, technology, and history.[9] The covering over of its messy interior is part of an American historical amnesia—believing it is possible to wipe almost any slate clean and start over. Yet this interior persists, bearing with it a history of relating to the land that does not fit within utopian pastoral landscape narratives.

More than three decades ago, the artist Robert Smithson revealed the sublimity of landscapes that were no longer pristine, suggesting a possible way to approach the paradoxical urbanity of Fresh Kills. Smithson's Nonsite projects represent a sublime in which repressed fears of contamination have replaced earlier cultures' repressed fears of nature (FIG. 6). These works of Smithson may be seen as a critique of the influence of a conventional notion of the sublime on the perception of American landscape: how the narrative of the sublime, in its idealization of America's exceptional place in the world, was a foil for expansionism, its characteristics of wildness, grandeur, and overwhelming power identified with manifestations of American nature, associated with the western frontier, as a way to promote or legitimate the violence of exploration and conquest, the appropriation of nature, and the displacement of native people.[10]

FIG. 7: Eighteenth-century landscape garden (top).
FIG. 8: Sublime American landscape (bottom).

In spite of the fact that the sublime seems to have been an embedded characteristic of the "new" world of America, inseparable from the awe-inspiring monumentality of its forms fading into the grandeur of its spaces, it arrived on this continent with its definition already consolidated, as a contradictory subtext within a pastoral aesthetic sensibility, as it had developed through the cultural artifact of the British landscape garden. On the one hand, as a representation of "wild" nature's potentially overwhelming power, the sublime needed to be domesticated; on the other, such domestication—in other words, the act of design—could not be acknowledged because of the desire to associate sublimity with a power greater than "man." Thus came about a fusion of a more or less transcendent sublime covertly contained within a formal aesthetic of the beautiful. This fusion produced the smooth naturalistic landscapes in which the "hand of man"—that is, design—was supposedly invisible (FIG. 7). Despite this degree of conceptual control, by the late eighteenth century the unacknowledged presence of the sublime within the landscape garden rendered its construction conceptually impossible, bringing the era of the English landscape garden to an end. Gardens became smaller and more obviously cultivated, their boundaries once again evident; "landscape" became instead the object of cultured tourism—first to England's Lake District, then further away, to America (FIG. 8). Travelers who journeyed to these places appropriated the language and the viewing techniques of the landscape garden to describe grand and wild "natural" prospects that opened up before their eyes.

The importation of the aesthetic language of landscape had a profound influence on the construction of American space, where there existed the simultaneous desire to control a powerful nature and to keep such control hidden. However, the scale shift from British landscape garden to American continent transformed what had been an agreed-upon aesthetic fiction of untouched nature in a garden into a national myth of an eternal return to a forever-unspoiled nature. Further, the "discovery" of America as the new "garden of the world" was (and continues to be) represented in terms of two spatially irreconcilable myths of origin and Golden Age: Eden, the bounded garden representing Judeo-Christian origins, and Arcadia (an inspiration for the landscape garden), the unbounded pastoral space of Greek shepherds and their flocks, a working landscape that does not appear to be a working landscape.[11] It is possible to see an ideological parallel between the perceived potential for "renewal" of the Fresh Kills site and the contradictory belief structure associated with the historical development of the American landscape that represents the blurring of pastoral and Edenic narratives in support of a utilitarian ideology. In theory, these are irreconcilable; in practice, the contradiction is disguised or ignored, producing an unworkable fusion, which has contributed to the deadlock of the pastoral park.[12]

The fear of nature associated with the imaginative symbolism of the sublime has to do with the fear of that which cannot be controlled. At Fresh Kills, the mounds of garbage have the potential to constitute a disturbance in the way that in a pre-industrialized period was the province of nature, including its size. Because the technology for addressing the mounds is based on covering them, they are closed forms. Such closure is characteristic of the aesthetic category of the beautiful. Yet it is possible to disturb their scenic unity to preclude the easy reabsorption of their internal disorder. In this context the act of design can itself be understood as a kind of agitation or disturbance, setting in motion forces and elements in ways that allow them to be apprehended.[13]

The sublime offers a means of addressing something too large or too complex to be comprehended—a provisional means of designing the indefinable. If a sublime space has the property of being uncontainable, it is not necessarily because it is transcendent but because it cannot be completely thought. It is an aesthetic based on the incommensurable. This identification with differences too large to be absorbed can be deployed to develop the multiplicity of identity of a large park.

Insofar as it is a historical construct, the sublime has been a way of coming to terms with infinity. Yet it is also a material space, albeit one that points outside itself. If it is possible to construct and represent the sublime, it is by displacing matter in such a way as to set up a movement around it, an invocation to something that is there but not containable. It is in part the materiality of the sublime that makes it relevant to the remaking of the Fresh Kills landscape, in terms of both the displacement that already occurred there and the further displacement that processes of design entail.

The sublime engages complexity, having to do with constructing an identity that is not singular. Its engagement of that which is uncontainable in time and space has a parallel in scales of ecological function that extend beyond the boundaries of the site, having to do, for example, with its occupation by migrating birds, traversing the east coast of the United States, or that exist as processes that occur over time, eluding a stable or static identity. There is a parallel between the sublime as an aesthetic addressing something too large or complex for the mind to grasp, and an ecological framework engaging forces and flows which are not contained or containable within a specific site.

Before Fresh Kills was a landfill, it was a wetland.[14] The relationship between Fresh Kills the closed and capped landfill and Fresh Kills the urban wetland is a crucial one in the context of creating a park on the site. Historically, landfills were often built on wetlands, which were considered dangerous, disease-carrying environments. They became used as dumps because they were perceived to be no better than dumps (FIG. 9). As Elizabeth Barlow Rogers wrote in 1969, "'Landfill,' when preceded by 'sanitary,' is a euphemism for 'garbage dump,' and the topography of marshes [now known as freshwater wetlands]—low-lying, level, treeless—makes them ideal municipal dumps, besides which, they often are already in the public domain."[15]

FIG. 9: View of Staten Island wetland as de facto dump, from 1930s.

The negative connotation attending the identity of the Fresh Kills site as landfill was already present in its historical identity as wetland. Such sites, commonly called swamps, were not only undesirable, they were also largely inaccessible, operating as huge barriers to different kinds of urban movement. Any design approach to Fresh Kills must address the inversion from undesirable space to public amenity. The fact that Fresh Kills was perceived as undesirable long before it became a landfill offers a key to engage this challenge, especially in that the portions of the site that have survived as wetlands now have a perceived positive value.

The transformation of wetlands from reviled swamp to appreciated natural area is supported by significant legislation, beginning with the Wetlands Protection Act of 1981. Because wetlands have only been protected from development since that time, most of those that still exist are located in underserved parts of cities. Robert Moses, creator of the Fresh Kills landfill, filled almost 2,000 acres of wetlands in New York City, in urban areas that had sufficient political representation to be deemed important enough for this service. Many wetland areas on Staten Island were filled after World War II so that they could be developed as much-needed housing, in the form of poorly planned and monotonous multifamily housing blocks and sprawling subdivisions. The wetlands that make up a third of the Fresh Kills site resisted this kind of development in part because of their proximity to the landfill.[16]

In the context of the shift in perceived value of wetlands from liability to asset, the transformation of this land, which was once a barrier, can provoke a rethinking of the urban as well as natural identities of the site. This process of rethinking can contribute, through the mechanism of environmental justice, to a cultural shift in the understanding of relationships between natural and social spaces.

As a messy, uncontainable, and potentially threatening side of nature, wetlands are not part of a pastoral landscape tradition. The recognition of their value can support the design of a park landscape that acknowledges disturbance, and as such is not wedded to the visual harmony that characterizes a pastoral park. "Landscape" in a pastoral pictorial sense is supposedly not dirty, because its dirt is sealed within a smooth surface that portrays a lack of disturbance. The porosity and mutability of a wetland ecosystem is threatening to the temporal and spatial stability of this pictoral identity.

Inseparable from the inversion of value of wetlands is the contemporary paradigm of non-equilibrium ecology that reframes nature in terms of its continual disturbance, rejecting the previous scientific "truth" of organic nature's tendency toward either equilibrium or homogeneity. The discovery that disturbance is fundamental to natural systems has provoked a shift away from the model of a single ecosystem to one of multiple systems overlapping in "a landscape of patches...responding to an unceasing barrage of perturbations."[17] As Steward T. A. Pickett and P. S. White describe:

> Ecologists have always been aware of the importance of natural dynamics in ecosystems, but historically, the focus has been on successional development of equilibrium communities.... Recently many ecologists have turned their attention to processes of disturbance themselves and to the evolutionary significance of such events.... The phrase "patch dynamics" describes their common focus.... Equilibrium landscapes would...seem to be the exception, rather than the rule.[18]

The significance of this paradigm shift in landscape ecological thinking and practice is apparent in Nina-Marie Lister's discussion of the difference between conventional and emerging ecosystem characteristics. A crucial point that Lister makes about the design and development of landscapes is that understanding comes from multiple perspectives, system types, and scales. Because of this combination of characteristics, decision making in the planning, design, and management of a landscape must be adaptive.

Contemporary ecology's investment in dynamic complex systems has a parallel in urban design, in terms of the interaction of social phenomena, in a field of tensions that includes multiple constituencies, forces, systems, and agencies. How is it possible to engage this multiplicity without containing it?

In 1967, as he was creating the Nonsite projects, Robert Smithson wrote:

> The artist must come out of the isolation of galleries and provide a concrete consciousness for the present as it really exists,...must accept and enter into real problems that confront the ecologist and industrialist.... We should begin to develop an art education based on relations to specific sites. How we *see* things and places is...primary.[19]

The artist's prescient words describe a connection between ecology and a post-industrial landscape, engaging the necessity for artistic representation (how we see things and places) of specificity, and of specific sites. His integration of materiality and site specificity in the act of representation is evident in his definition of the Nonsite as "a changing reality with a material basis."[20] He linked this changing reality to the notion of entropy as a primary aspect of the dynamic identity of this post-industrial landscape. *A Nonsite, Pine Barrens, New Jersey*, the artist's first Nonsite, constructed in 1968, includes an aluminum bin holding sand, an aerial photograph, and a map. (The Fresh Kills site includes a similar pine-barrens ecological system.) Other Nonsites represent post-industrial landscapes such as mines, with steel and slag included in the gallery installations. Smithson was looking for "sites that had been in some way disrupted...pulverized...for denaturalization rather than scenic beauty."[21]

In the Nonsite sculptures, Smithson defined the gallery—operating as a stand-in for the city—as the second site of the work. The first site, the site of extraction, was the disused industrial area, which he documented through photographs, maps, and material samples to construct the Nonsite. The identity of the work is produced through an oscillation that results from a dynamic process of reconnecting materials that the artist has displaced from each other. This kind of oscillation can set up a movement around a site to construct a sublime space.

Yet there is a potential paradox in this comparison. For Smithson, the city-gallery was the second site. In the case of Fresh Kills, the city is the "first" site, from which materials have been displaced. The landfill occupies the position of the gallery, in the sense that it is the second site, to which materials have been brought. Yet Fresh Kills occupies a geographical position similar to that of Smithson's "first" sites. Many of these sites were in New Jersey, some near Fresh Kills, at a similar degree of remoteness from Manhattan, where the work was exhibited. As the Fresh Kills site is reconstructed as a park, its historical identity as a landfill will also be represented as part of the design. As such, it will operate as both first and second site—site of extraction and site of representation.

The parallel between Fresh Kills and the Nonsites has to do not only with the displacement of materials from one site to another but also with the representation of identity of a site. Smithson described them as "gallery sculptures...containing bins, rocks, and maps that refer back to the Site."[22] A work such as *Six Stops on a Section* is a matrixlike array in which multiple forms of documentation interact dynamically to represent a site, without pictorializing it or allowing it to be reduced to any one thing (FIG. 6). This composite representation of the site is an important contribution of the Nonsite series in reference to the construction of identity in the large park.

With the Nonsites, Smithson demonstrated that landscape—as nature, and thus as traditional inspiration for art—is no longer pure. Each Nonsite is

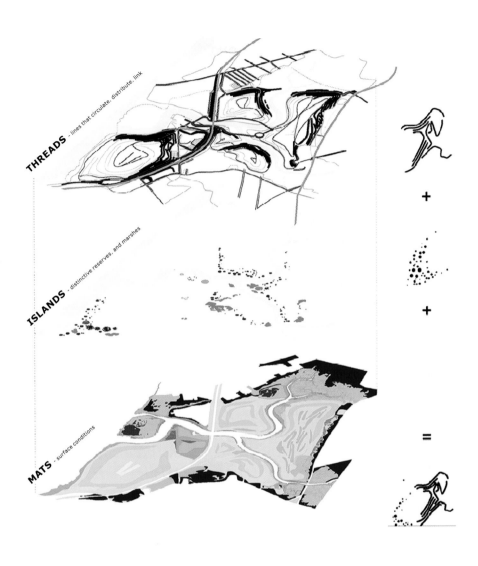

FIG. 10: James Corner/Field Operations et al., "Lifescape," Fresh Kills competition proposal, diagram showing systems of threads, islands, and mats.

HABITAT

PROGRAM

CIRCULATION

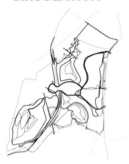

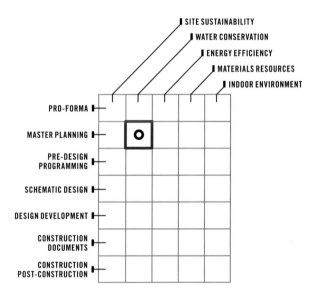

FIG. 11: James Corner/Field Operations et al., Fresh Kills Park, schematic design diagram showing systems of habitat, program, and circulation (top).
FIG. 12: "Green matrix" (bottom).

a "fragment of a greater fragmentation."[23] Urban desires and trajectories have not only "touched" nature, they have irrevocably altered its identity. This alteration challenges an essentialist polarizing of the terms *city* and *nature*, each with its own stable identity, suggesting instead a more historically layered and internally differentiated conception of identity.

Keeping track of and orchestrating such dynamic complexity requires conceptual tools. A matrix, in the mathematical sense, is a tool for dealing with complexity. The use of this type of matrix to array and calibrate relationships between parameters in a dynamic fashion can allow the park to be more than a collection of programs on the one hand, or a naturalistic preserve on the other. In line with Lister's model of adaptive design, this layering of parameters requires and enables thinking provisionally, for instance in terms of scenarios, such that unanticipated spatial characteristics may emerge from the interplay between elements.

In mathematics, a matrix is understood as a rectangular table of elements, considered as an entity, which may express quantities or operations. As opposed to a grid, which can hold many kinds of information but does not interact with its contents or set up such content to interact with itself, a matrix is conceived in dynamic terms: engaging the action of each term on others in what becomes a set of dynamic relationships.

A mathematical matrix is useful to keep track of data that depend on several parameters. In mathematics, these data would be coefficients of systems of linear equations and transformations. An example of a digital matrix used to manage references and resources in relation to phases of project design is the so-called "green matrix," a website that enables access to resources on sustainable design (FIG. 12).[24] Across the horizontal axis of the matrix are sustainability topics such as water conservation, indoor environment, and energy efficiency. The vertical axis reflects phases of project design, such as master planning, schematic design, and construction drawings. A user places his or her cursor at the intersection of a topic and a phase to find relevant resources.

In a design project, a matrix can support the construction of a kind of unity that does not rely on a single vision or overarching order to manage in creative and operational terms the interactions between multiple perspectives, scales, and types that attend the development of a complex urban ecological landscape such as Fresh Kills. This type of matrix supports thinking in terms of sustainability: not only in regard to a project's changing future, but also because such a project cannot be conceived as starting with a clean slate. It is necessary to represent what is already there, in terms of engineering processes and artifacts as well as existing ecologies, so that it can be considered in the mix, to design something that cannot be fully understood from the beginning.

A matrix can acknowledge relationships between large and small scales operating simultaneously in a site, to allow the making and interweaving of radically different kinds of places. It can enable a language of ecological systems as discussed above to take form in landscape architecture, to develop frameworks that are simultaneously mutable and precise, which can embrace spatial and temporal complexity within the structure of a project.

Three out of the six proposals for the Fresh Kills competition utilize some kind of mathematical matrix to represent the project. Structuring their competition projects according to a matrix of operations and/or elements, which interact with each other, in support of a strategy of conceptualizing the park, allows the proposals of Hargreaves Associates, Field Operations, and Mathur/da Cunha + Tom Leader Studio to articulate hybrid strategies that construct an array of temporal and spatial, social and natural qualities in dynamic relationships that include the site in its complex internally differentiated identity.

Looking at these proposals and at the development of the Lifescape project can help to delineate meanings and uses of the concept of matrix as a support through which to address identity in dynamic terms in the design of a large park. Without delving into matrix theory, it is possible to perceive that the potential for dynamic identity that may emerge from mutually dependent relationships is relevant to the understanding of these projects—such that their use of a matrix goes beyond metaphor.

Each of these proposals, by acknowledging the possibility that different aspects of a large and complex site operate according to divergent trajectories, implies that different aspects of a project might also operate in such a way, therefore resisting any organic integration between differences. Tensions will always exist in design between past and future, protection and access, built and natural areas. A design approach that recognizes such tensions can deploy them fruitfully. A matrix can support the mapping of tensions between different kinds of operations and orders that within a singular convention of representation could not be seen, to bring them to the surface within a site, without having to either resolve or remove them, alternatives that are in some cases impossible. The use of a matrix can help to conceptualize an intervention in terms of an apparatus that "captures the momentary overlap, interface, and resonance among elements that create a field of intensification of existing conditions."[25]

The Mathur/da Cunha + Tom Leader Studio proposal uses an elaborate mathematical type of matrix to construct parallels between abstract and material aspects of the historical composition of the site and its remaking. Five spatial terms—surface, field, datum, edge, and zone—are aligned with five radically different kinds of materials, respectively: 1) World Trade Center debris (as closest to the surface of the landfill); 2) household garbage (as the most generalized, that is, field condition); 3) *Spartina*, the marshland grass (whose water level the proposal posits as a new datum); 4) glacial till (reflecting the site's location as the farthest edge of the glacier that retreated across

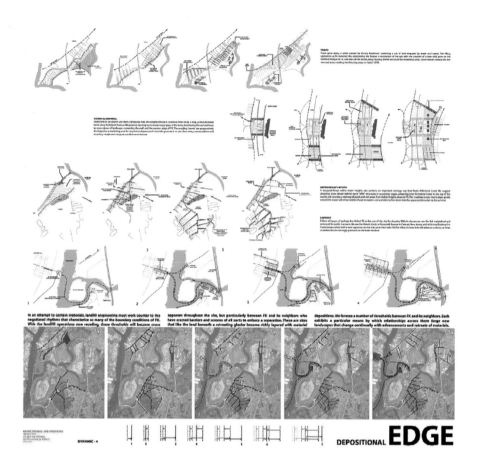

FIG. 13: Mathur/da Cunha + Tom Leader Studio et al., "Dynamic Coalition," Fresh Kills competition proposal, diagram of edge relationships.

the island millennia ago); and 5) rocks occupying a fault zone running deep in the ground beneath the site. Each of the five spatial terms is also aligned with a proposed action, which is grounded in landscape or urban ecological strategies; to construct a threshold that addresses the park edge as a social landscape, for instance, gaps are opened in the forested mound separating Fresh Kills from the Staten Island Mall, to create market areas for recycled goods (FIG. 13).

Mathur/da Cunha + Tom Leader Studio matrix's emphasis on performance rather than form enables a consistency of approach toward materials that otherwise would remain locked in conventional categories of historical or contemporary, artificial or natural, pure or impure, and therefore only approachable on vastly different terms. The proposal recasts the site as a landscape made up of fragments of different times and places, projected into the future. Deriving its logic and aesthetic from an understanding of the site as an accumulation of heterogeneous actions over time allows the proposal to explicitly include the waste materials of the landfill.

Matrix is a term with multiple meanings. George Hargreaves, in this volume, begins his discussion of large parks with a definition of matrix as a "mold in which a thing is cast or shaped" to articulate the necessity of having disparate forms and site conditions within one design in order to sustain the complexity of a large park. This use of the term reflects its usage in landscape ecology, where it is a fundamental component of landscape structure. As the dominant component in a landscape, the matrix is the most extensive and connected landscape type. If landscape ecologists try to manage a habitat without considering the matrix, they will likely fail to provide what wildlife need in that area.

Hargreaves uses the concept of matrix to enable a different perception of spaces already designed and inhabited—a tool through which to see historical landscapes differently. For instance, in the Amsterdam Bos Park, he asserts that the presence of two matrices overlaid with a management plan that does not try to reconcile them is what "imbues the park with its experiential qualities." In a critique of landscape architectural practice, he states that "slavery to oneness, the unified plan, the need to remake an entire site, and abhorrence of program keeps [landscape architects] from realizing the full complexity and diversity that parks can have."[26]

In landscape ecology, characteristics of matrix structure include porosity (or the density of patches) boundary shape, networks, and heterogeneity. Within matrix areas, networks connect habitats of different size and shape, supporting heterogeneity within the landscape. The fact that different habitat patches usually are replicated throughout the matrix suggests a way of thinking about programming and occupying a large site.

Richard Forman and Michel Godron devote a chapter of their textbook on landscape ecology to "the challenge of understanding the nature of the landscape matrix."[27] As they describe it, the matrix plays the dominant role in the functioning of the landscape, including the flows of energy, materials, and species.[28] At one extreme, "some landscapes have an extensive homogeneous matrix containing scattered distinct patches"; at another, "an entire landscape may be composed of small patches that differ from one another."[29]

Forman and Godron refer to three definitions of matrix used in diverse areas of knowledge: 1) the homogeneous mass in which small differentiated elements appear; 2) the binding material that surrounds and cements independent elements; and 3) the mold in which a metal sculpture is produced or, in vertebrates, the organ in which an embryo develops. For each of these three definitions, they articulate formal criteria for understanding a landscape. For instance, the first definition of matrix as homogeneous mass relates primarily to area: size or scale is an index of the role of a matrix in a landscape in that "generally, the area of the matrix exceeds the total area of any other landscape element type present."[30] The second definition of matrix as a binding material furthers the understanding of the role of the landscape matrix in supporting heterogeneity.

One definition of a matrix that seems to unite several other meanings is that in which the term is understood as "a situation or surrounding substance within which something else originates, develops, or is contained."[31] It comes from the Latin word for "womb," which is derived from *mater*, the Latin word for mother. This sense of a matrix is both formative and generative. It also exists in the definition of the term as "something that constitutes the place or point from which something else originates, takes form, or develops, as in 'The Greco-Roman world was the matrix for Western civilization.'"[32] This sense of the term makes it possible to conceive of a design project in terms of setting a matrix in place.

The Field Operations and the Mathur/da Cunha + Tom Leader Studio competition proposals are examples of how such a framework can bring the dimension of time into the construction of a park, recognizing the complex phenomenon of emergence. In these projects, a matrix provides a way of engaging radically different kinds of information: it can extend the ordering and design of a landscape beyond what are conventionally accepted as ecological systems into explicitly social and urban relationships.

The Hargreaves team's proposal for "Parklands" uses the concept of a landscape matrix as a support for heterogeneity at different scales: three linked landscape strategies—transformation, succession, and operation—emphasize the site's role as a living, dynamic entity. Each strategy has a temporal component, which provides a basis for the ecological development of the site. This ecological development in turn supports the spatial construction of the site as three conceptually overlapping environments: the Domain, which includes a significant amount of vegetation, but is understood primarily as social space, whether individual or collective, indoors or out; the Meadows, which focuses on natural phenomena in a way that registers human perception of these phenomena; and Lake Island, a bird sanctuary on a newly reconstructed island that marks a new center for the park as a habitat (FIG. 14).

According to Hargreaves's terminology, each of the three environments of the Parklands proposal constitutes a landscape matrix, as the dominant landscape type in a particular area. The fact that each one conceptually overlaps the other two, as with a Venn diagram, establishes the potential for an overall unity that does not rely on a singular identity. The proposal attains an economy of means by establishing partial similarities between different environments: areas for people in support of nature; areas focusing on nature to engage people; the interdependence of architecture with existing infrastructure in support of social and natural spaces to engage multiple scales of the site without necessitating a large quantity of building. The primary architectural intervention, a Hall of Exhibits and Inquiry, is sited alongside the bridge across the wetland, at the main route through the site.

Field Operations' winning Lifescape project is described on the competition boards as a "reconstituted matrix of diverse life forms and evolving strategies." This matrix supports the integration of physical design with geological, hydrological, and biological processes at multiple scales. The proposal's spatial framework of threads, islands/clusters, and mats can be understood as the agent of a fluid set of ecological systems, allowing the interaction of programmatic, cultural, and natural elements to create the complex, synthetic environment (FIGS. 10, 11). In ecological terms, for example, Fresh Kills Park will support diverse habitats including salt marsh, native prairies, maritime oak forest, birch thickets, and pine/oak barrens (FIG. 15).

Each of the three sets of elements—threads, islands, and mats—operates as a strategy, a figure, and a device of representation. Each set of elements constitutes a matrix in a landscape-ecological sense, when it becomes a dominant landscape type. Each matrix has its own rules of organization and its own variables, which don't necessarily relate to those of the other two. The proposal groups all kinds of things (plantings, activities, buildings) according to the three matrices, using them to cut across categories to construct new identifications, based on internal differences, to support a kind of "precise openness," which offers a potential for change.[33]

The thread element in Field Operations' proposal for the phased densification of the landscape of the mounds operates according to the logic of a hedgerow network. As described by Forman and Godron, a hedgerow network engages a further criterion for determining a matrix. A hedgerow may be highly dynamic when made up of pioneer successional species, so that it can act as a species source to initiate a future landscape. In spite of the fact that a network of hedgerows generally covers less than a tenth of a landscape's total area, it may be understood as its matrix, in that it encloses the other parts of the terrain. A hedgerow fulfills the definition of a matrix as "the binding that surrounds and cements the independent elements."[34]

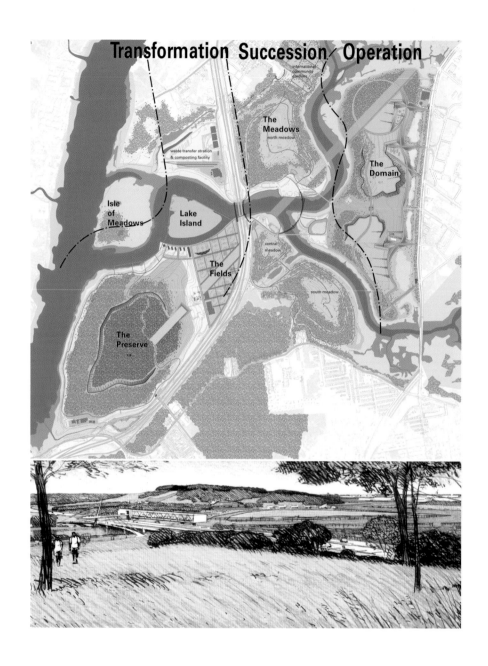

FIG. 14: Hargreaves Associates et al., "Parklands," Fresh Kills competition proposal, plan showing three landscape matrices, and view of parklands.

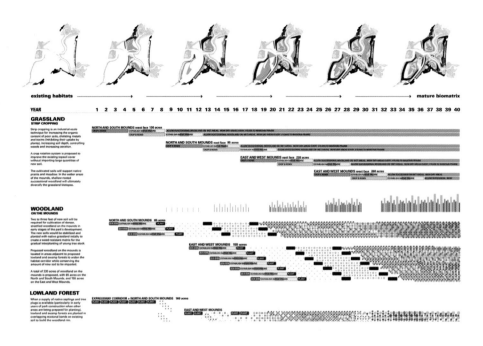

existing habitats ————————————————————————————————→ mature biomatrix

YEAR   1 2 3 4 5 6 7 8 9 10 11 12 13 14 15 16 17 18 19 20 21 22 23 24 25 26 27 28 29 30 31 32 33 34 35 36 37 38 39 40

GRASSLAND
STRIP CROPPING

Strip cropping is an industrial-scale technique for increasing the organic content of poor soils, chelating metals and toxins (inhibiting their uptake by plants), increasing soil depth, controlling weeds and increasing aeration.

A crop rotation system is proposed to improve the existing topsoil cover without importing large quantities of new soil.

The cultivated soils will support native prairie and meadow. In the wetter areas of the mounds, shallow-rooted successional woodland will ultimately diversify the grassland biotopos.

WOODLAND
ON THE MOUNDS

Two to three feet of new soil will be required for cultivation of dense, stratified woodland on the mounds in early stages of the park's development. The new soils would be stabilized and planted with native grassland initially to create a weed-resistant matrix for the gradual interplanting of young tree stock.

Proposed woodland on the mounds is located in areas adjacent to proposed lowland and swamp forests to widen the habitat corridor while conserving the amount of new soil to be imported.

A total of 220 acres of woodland on the mounds is proposed, with 65 acres on the North and South Mounds, and 155 acres on the East and West Mounds.

LOWLAND FOREST

When a supply of native saplings and tree plugs is available (particularly in early years of park construction when other areas are being prepared for planting), lowland and swamp forests are planted in overlapping ecotonal bands on existing soil to build the woodland rim.

FIG. 15: James Corner/Field Operations et al., Fresh Kills Park, diagram of base process, 2005.

phasing and development sequence

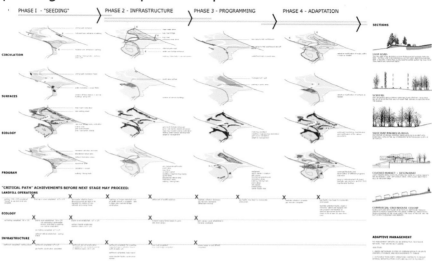

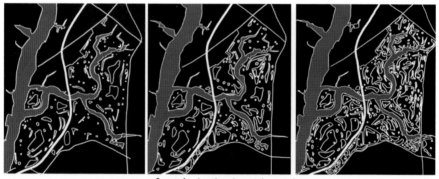

a process of recolonization is set in motion ...

FIGS. 16, 17: James Corner/Field Operations et al., "Lifescape," Fresh Kills competition proposal, diagram of phasing and development (top) and growing a park, 2005 (bottom).

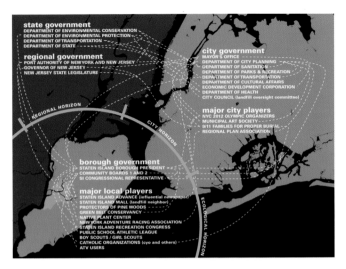

FIG. 18: James Corner/Field Operations et al., mapping of agencies involved in Fresh Kills project, 2005.

In Lifescape, the thread element is used to locate and organize trees in a way that exploits the terraced contouring of the mounds' final cover. The greater depth of soil at the outer edge of each bench supports the trees' root systems, while at the same time the swale at the inner portion of the bench collects water for irrigation. Locating the planting on the northeast sides of the mounds takes advantage of and makes visible a microclimate to establish the new environment, avoiding the prevailing winds from the southwest (FIGS. 16–18).

Corner's role and profile in a larger debate about the field of landscape architecture is part of what has allowed the Fresh Kills project to gain credibility, to adjust its own rhetoric in the context of the years of public process and approvals since the competition. The restructuring of the balance of components of the project throughout this public process is a key part of the development of a park of this size and scope, which Field Operations has conceptually layered onto other kinds of processes.

A key aspect of the transformation of landfill to landscape is the enabling of new flows in relation to human occupation along the edges of the site. Like other brownfield sites, as discussed above, the edges of a landfill tend to be dysfunctional places—often a mass of dead-end streets and backs of buildings—partially a consequence of their adjacency to an undesirable location. Yet these edges have a significant potential, as new thresholds, to engage and leverage a host of urban concerns alongside the development of a park. To formulate a strategy for edges, it is necessary to recognize how the social and natural histories of the site at various scales have produced its differences from its surroundings. As part of the public process since the competition,

LAYERS OF FRESH KILLS lifescape

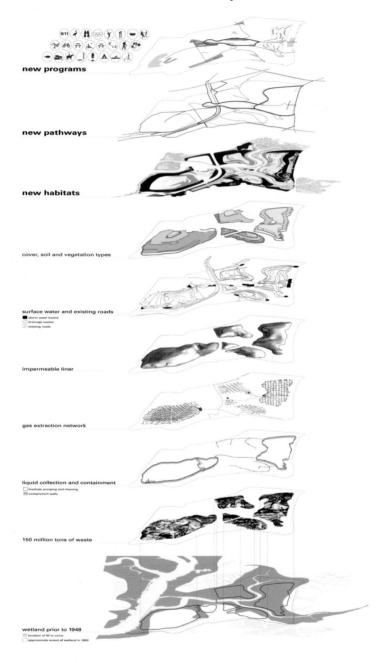

new programs

new pathways

new habitats

cover, soil and vegetation types

surface water and existing roads

impermeable liner

gas extraction network

liquid collection and containment

150 million tons of waste

wetland prior to 1948

FIG. 19: James Corner/Field Operations et al., "Lifescape," Fresh Kills competition proposal, spatial framework: layering of three new systems onto existing site systems.

Field Operations has developed a strategy of neighborhood parks, located at the edges of the site, within the larger Fresh Kills Park.

From the beginning, Field Operations has grounded the identity of Lifescape in a matrix that takes processes of realization of the project into account. The project for Fresh Kills must calibrate and orchestrate the site's complexity as a brownfield in temporal as well as spatial terms. Phasing is a significant engineering and ecological challenge, factoring in approximately thirty years for garbage decomposition, methane production, leachate drainage, and up to one hundred feet of differential settlement of the mounds. The more precise such orchestration, the greater opportunities for phasing to enrich and enable the project.

Field Operations has outlined a long-term strategy based on natural processes, agricultural practices, and plant lifecycles to rehabilitate and transform the degraded site over the next thirty years. This strategy has evolved from the competition entry, in which a matrix addressed six project phases, each with discrete stages, through which the park would be "grown," as in seeding, cultivating, propagating, and evolving. As Corner has stated, design at Fresh Kills is as much about the "design of a method and process of transformation as it is about the design of specific places."[35] As the project has evolved, its spatial framework has been cast in terms of four phases: seeding, infrastructure, programming, and adaptation. The first three phases are the main development over a thirty-year time frame. The adaptation phase reserves further possibilities for negotiations to respond to changing needs and circumstances.

As in the competition entry, three organizational systems construct the formal, material, and programmatic fabric of the overall Fresh Kills landscape in the current project. The identity of these systems, however, has changed in response to processes of project development. The first is a matlike surface of grasses, whose grain of pattern is finer in areas of higher use, coarser in more passive areas. The same grain, although not the same material, governs the treatment of paved areas. The second system—threads—has become circulation loops, including existing and new roads, paths, and trails that allow for movement around the site while completing circuits so that visitors can end up where they started. The third system has evolved from island to forest: woodland that overlays the landscape, mixing different ages of mostly native planting. Exotic and unusual species are concentrated in cloudlike arrangements in certain areas (FIG. 19).

There is a difference between Field Operations' approach and that of James Corner's teacher Ian McHarg, whose *Design with Nature* introduced a method of landscape analysis that has contributed to an understanding of the layering of different parameters in the design of a landscape.[36] The flaw in McHarg's method was his use of such analysis as a deferral or substitution for design issues. Field Operations' approach goes beyond problem solving, while using some of the same techniques of representation.

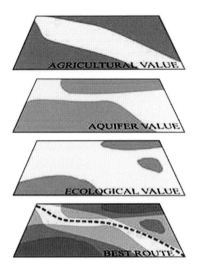

FIG. 20: Ian McHarg's overlay method to evaluate positioning of road in Staten Island.

One reason why environmental considerations played a small role in planning and design prior to McHarg was the lack of a method to represent them. His use of map overlays conveying large amounts of spatial information in a concise manner was a forerunner to forms of complex analysis of multiple criteria in evaluation and decision making. *Design with Nature* reproduces his analysis of where to situate a major new road in Staten Island with respect to "social values," in terms of benefits and costs to society, engaging factors of historic, water, forest, wildlife, scenic, recreation, residential, institutional, and land values. Each factor or value is represented by a map transparency, with the darkest tones representing areas of greatest value. Superimposition of the transparencies over the original map showed the darkest areas as having the greatest overall social values, and the lightest as having the least. The social-value composite map was then compared with similar maps of geologic and other considerations. The result of these comparisons would determine where to situate the road (FIG. 20).

The transformation of a matrix from a tool for classification of complexity into a springboard for figuration happens in the case of Field Operations' design for Fresh Kills through multiple techniques of representation, with an emphasis on collage, producing a field in tension that activates components that emerge from matrix thinking. The inventory of components or site systems are represented in images in which the tension between the parts sustains the charge of their individual connotations. This approach activates imaging in a way that is matrix-based, in that the relationship between those components is understood dynamically through the multiple points of engagement that a collage construction of space enables.

0 - 15 YEARS
HABITAT DIVERSIFICATION OVER TIME
early stages: preliminary plantings related to
existing biomass and habitat

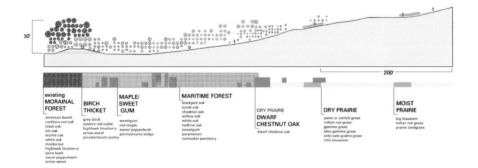

existing MORAINAL FOREST	BIRCH THICKET	MAPLE/ SWEET GUM	MARITIME FOREST	DRY PRAIRIE DWARF CHESTNUT OAK	DRY PRAIRIE	MOIST PRAIRIE
american beech northern red oak black oak pin oak scarlet oak white oak mockernut highbush blueberry spice bush sweet pepperbush arrow wood	grey birch eastern red cedar highbush blueberry arrow wood pinxsterboom azalea	sweetgum red maple sweet pepperbush pennsylvania sedge	blackjack oak scrub oak chestnut oak willow oak white oak rudkins oak sweetgum persimmon nantucket juneberry	dwarf chestnut oak	panic or switch grass indian nut grass gamma grass blue gamma grass side-oats grama grass little bluestem	big bluestem indian nut grass prairie cordgrass

15 - 30 YEARS
HABITAT DIVERSIFICATION OVER TIME
developed stages: overlapping inter-plantings and
"spread" of seed bank and species, establishing
stratified habitat communities and diverse
ecological matrices

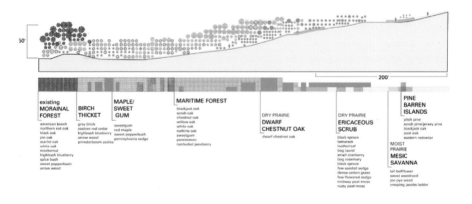

existing MORAINAL FOREST	BIRCH THICKET	MAPLE/ SWEET GUM	MARITIME FOREST	DRY PRAIRIE DWARF CHESTNUT OAK	DRY PRAIRIE ERICACEOUS SCRUB	PINE BARREN ISLANDS MOIST PRAIRIE MESIC SAVANNA
american beech northern red oak black oak pin oak scarlet oak white oak mockernut highbush blueberry spice bush sweet pepperbush arrow wood	grey birch eastern red cedar highbush blueberry arrow wood pinxsterboom azalea	sweetgum red maple sweet pepperbush pennsylvania sedge	blackjack oak scrub oak chestnut oak willow oak white oak rudkins oak sweetgum persimmon nantucket juneberry	dwarf chestnut oak	black spruce tamarack leatherleaf bog laurel small cranberry bog rosemary black spruce few-seeded sedge dense cotton grass few-flowered sedge midway peat moss rusty peat moss	pitch pine scrub pine/jersey pine blackjack oak post oak eastern redcedar tall bellflower sweet woodrush joe-pye weed creeping jacobs ladder

FIG. 21: James Corner/Field Operations et al., sections of Fresh Kills Park showing habitat diversification
over time.

Collage-based techniques can recognize a site's existing elements and histories, as a way of entering into an already urbanized situation rather than a fictional clean slate. Each element has its own force and dynamic, which can unfold on its own spatial and temporal terms. In other words, it thickens the plane of representation through this overlap, which also includes different temporalities.

Lifescape's mats, threads, and forests are material. A matrix that is material as well as diagrammatic provides a way of approaching the design of the ground that acknowledges its potential complexity—that is, its internal differentiation—precluding a perception of ground as singular origin or essentialized nature. The term "matrix" can substitute for the term "ground" in a way that invests ground with a multiplicity and a resonance that precludes its collapse back into an unquestioned background. As well, matrix has a simultaneously constructed and natural sense. It cannot be reduced to inert background, in part because it explicitly includes processes, including processes of gestation (FIGS. 21, 22). As the phasing diagrams make clear, the different ecosystems of the site will emerge and become legible in time.

It is materiality that provides a point of intersection between the notions of the sublime and the use of a matrix in the design of a project. Materiality in this case refers to living systems, including those of the operations of the landfill. In this context, materiality cannot be understood as the stable outcome of a process of forming or forging on the one hand, or as the stable referent to an act of perception on the other. The notion of materiality is reinscribed in larger processes, not only in terms of space but also in terms of time, including entropy, whereby a material manifestation is related to cycles of longer time frames and a condition of consistency that includes processes of decay as well as growth. Materials have conventionally been looked at as inanimate either because they have been extracted from a cycle of nature or because they are the byproduct of processes of production. This reconsideration of materiality has to do with a different perception of time, including its flows, directions, and rhythms. In the context of a subject-object relationship, it means that the object is not stable, because it too is part of processes of growth and decay.

FIG. 22: James Corner/Field Operations et al., aerial view of Fresh Kills Park.

NOTES

1. Although nothing in the competition brief explicitly defined the project as a park, its complexity as a building site combined with adjacent communities' resistance to other programs made a park the program of choice.

2. New York City Department of City Planning, "Fresh Kills Lifescape," http://www.nyc.gov/html/dcp/html/fkl/ (accessed June 11, 2006).

3. Fresh Kills' complexity as a site is apparent in the project website, for instance, in the glossary of more than two hundred terms. "Complexity" may be understood as the degree to which a system has a design or implementation that is difficult to understand and verify. The Latin word *complexus* signifies entwined or twisted together. This may be interpreted in the following way: to have a complex, you need two or more components, which are joined in such a way that it is difficult to separate them, in other words, something that is both intricate and compounded. The *Oxford English Dictionary* defines something as "complex" if it is "made of (usually several) closely connected parts," which are at the same time distinct and connected. A system is more complex if more parts can be distinguished, and if more connections exist between them. Whether a system is biological, technical, social, or otherwise, more parts to be represented means more extensive models, which require more time to be developed.

4. This essay continues an exploration of Fresh Kills begun as "Sublime Matters," published in *Praxis: Journal of Writing and Building,* Issue 4: Landscape (2002). Where the earlier essay focused on the competition, this text looks at Fresh Kills specifically as a large park—that is, in terms of its size and complexity—in relation to design frameworks of approach and realization. It also looks at the development of the Lifescape project, in operational terms, in regard to how to orchestrate a project that engages a level of complexity such as that inherent in the development of the Fresh Kills landscape.

5. The footprint of each mound is made up of two "sections." Each section was at one time an individual mound of garbage. The four mounds, in addition to being identified by numbers, are now identified as north, south, east, and west, with 1/9 being the west mound.

6. Brownfields are properties that are underutilized or abandoned due to environmental contamination associated with past industrial or waste-disposal use. Since the 1990s, federal and state programs have sought to encourage both private- and public-sector redevelopment of brownfield sites. See Thomas Russ, *Redeveloping Brownfields* (New York: McGraw-Hill, 2000). The most well known of such park projects is Landschaftspark Duisburg-Nord by Peter Latz + Partner. The sites of Downsview Park and Parc de la Villette are also brownfields.

7. Laderman Ukeles's three-screen video installation was shown simultaneously at Snug Harbor Cultural Center and the Staten Island Mall in 2002.

8. New York City Department of City Planning, http://www.nyc.gov/html/dcp/html/Fresh Kills/ada/competition (accessed June 11, 2006).

9. In fact, the "cap and cover" reclamation process is not one of renewal but of revegetation, in which the buried garbage gradually becomes inert, but does not biodegrade.

10. Elizabeth K. Meyer, in an essay entitled "Seized by Sublime Sentiments: Between *Terra Firma* and *Terra Incognita*," presents an argument for the sublime as a contemporary framework for interpretation, in the midst of the echoes of past disturbances, in William S. Saunders, ed., *Richard Haag: Bloedel Reserve and Gasworks Park* (New York: Princeton Architectural Press, 1997).

11. Leo Marx, "The American Ideology of Space," in Stuart Wrede and William Howard Adams, eds., *Denatured Visions: Landscape and Culture in the Twentieth Century* (New York: Museum of Modern Art, 1991), 62–78.

12. In the language of psychology, fusion refers to a dysfunctional state in which identities of two individuals merge, with the consequent blurring and loss of individual identity. To draw a parallel in spatial terms, it is desirable to sustain the combination of elements in a way that does not result in such blurring.

13. See Linda Pollak, "American Ground: Four Kinds of Disturbances," *Lotus* 100 (1998): 104–28, for discussion of relationships between design and disturbance.

14. Fresh Kills is also the name of the creek that runs through the site. "Kill" means "stream" in Dutch.

15. Elizabeth Barlow Rogers, *The Forests and Wetlands of New York City* (Boston: Little, Brown, 1969), 36.

16. While nearly 45 percent of the site was used for landfill operations, the remainder of the site is composed of wetlands, open waterways, and unfilled lowland areas.

17. Donald Worster, *The Wealth of Nature: Environmental History and the Ecological Imagination* (New York: Oxford University Press, 1993), 164.

18. Steward T. A. Pickett and P. S. White, *The Ecology of Natural Disturbance and Patch Dynamics* (Orlando,

FL: Academic Press, 1985), xiii, 5, 12.

19. Nancy Holt, ed., *The Writings of Robert Smithson* (New York: New York University Press, 1979), 221.

20. Ibid., 172.

21. Ibid.

22. Robert Hobbs, ed., *Robert Smithson: Sculpture* (Ithaca, NY: Johnson Museum of Art, 1980), Introduction.

23. Holt, *The Writings of Robert Smithson*, 84.

24. Ratcliff, "Ratcliff Green Matrix," http://www.greenmatrix.net (accessed June 6, 2006).

25. Sandro Marpillero, conversation with the author, May 3, 2006.

26. George Hargreaves, "Large Parks: A Designer's Perspective," in this volume.

27. Richard T. T. Forman and Michel Godron, "Matrix and Network," in *Landscape Ecology* (New York: Wiley, 1986), 157. The authors reference an earlier text by A. W. Kuchler, "Natural and Cultural Vegetation," *Professional Geographer* 21 (1969): 383–85.

28. Forman and Godron, "Matrix and Network," 159.

29. Ibid., 157.

30. Ibid., 161.

31. Webster's Online Dictionary, "Matrix," http://www.websters-online-dictionary.org/definition/matrix (accessed June 6, 2006).

32. Ibid.

33. Anita Berrizbeitia, "Scales of Undecidability," in Julia Czerniak, ed., *Downsview Park Toronto* (Munich and Cambridge, MA: Prestel and the Harvard University Graduate School of Design, 2002), 116–25.

34. Forman and Godron, "Matrix and Network," 164–65.

35. Field Operations project outline, Fresh Kills website, http://www.nyc.gov/html/fkl/ada/competition/2_3.html (accessed May 24, 2006).

36. Ian McHarg, *Design with Nature* (Garden City, NY: Natural History Press, 1969).

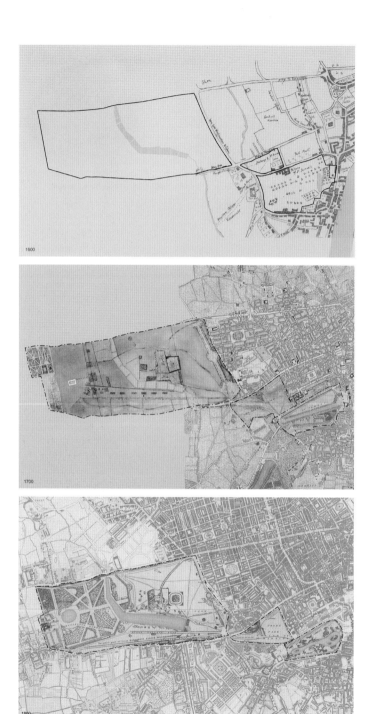

FIG. 1: Evolution of Hyde Park, Green Park, St. James Park, and Kensington Gardens, 2003.

LARGE PARKS:
A DESIGNER'S PERSPECTIVE

GEORGE HARGREAVES

Why large parks? Why does the history of large parks matter? What leads the public to embrace a landscape or park as its own? What kinds of sites and designs generate the fiscal resources necessary for their upkeep and very survival? Are there strategies that attenuate the ebb and flow of political regimes? These questions led me as a designer and builder to the great projects of history, not for mimetic pursuits but to identify underlying currents conducive to longevity, permanence, and importance within the culture.[1] As to the question of "large," well…size matters. The large park—in this case, one greater than 500 acres—affords the scale to realistically evaluate the degrees of success or failure of many of the issues embedded in public landscapes: ecology, habitat, human use and agency, cultural meaning, and iconographic import, to name but a few. These issues cannot be understood without considering the physical characteristics of the site itself, the fundamental base of any landscape. Large parks reveal the importance of the designer's attitude toward the site and its physical forms and natural systems. Site characteristics that could be masked or modified at a smaller scale are difficult to disguise and permanently transform across 500 acres. The extent to which designers embrace or fight the physical history and systems of a site is an important determinant of a park's long-term success.

Large parks raise questions about the design of large public spaces and their sites that are relevant to landscape architecture practice today. In Hyde Park and the Bois de Boulogne, we find parks whose site characteristics continue to imbue each park with particular qualities, drawing on a long and rich history of forested land, available water, and royal hunting parks. Both the Hyde Park complex and the Bois de Boulogne came about largely because of their sites, and the sites' physical histories continue to permeate each park with its specific defining qualities. In today's environment, where few if any such magnificent park sites are left, how does a designer approach a site that has been made and remade? In Golden Gate Park in San Francisco and Centennial Parklands in Sydney, we find parks that remade, if not erased, their identifying site characteristics in the name of pursuing the public's prescribed image of what a park should look like. Golden Gate's sand dunes and Centennial's swamplands were replaced with the lawns and deciduous trees characteristic of New York City's Central Park or London's Hyde Park, and as a result unsustainable park systems were created. How do we achieve sustainability today in an environment where the public may demand something else?

Amsterdam's Bos Park and the Parc du Sausset, outside of Paris, illustrate the richness of designing with multiple matrices within one park, avoiding the unified plan and the need to remake an entire site. These parks layer and juxtapose the made and the unmade, the designed and the undesigned, raising the issue of process and the difference between operation and process.[2] This is particularly relevant today in the landscape architecture environment that addresses a reclaimed site. There is a debate about where remediation processes cease to be the determining factor of public experience and instead human issues of connectivity, measure, and event predominate. In the seven case studies that follow, of large parks I visited and photographed, we see the work of designers who have grappled with these issues, and we focus on the various sites, in particular on their scale and defining characteristics.

Hyde Park Complex: The Park as Palimpsest

As a group, Hyde, Green, St. James, and later Kensington Gardens form a collective public park visited by virtually every London tourist, resident, and landscape architect. One of the earliest public landscapes in the Western world, with public access dating back to the 1600s, it is a touchstone in any discussion of public landscapes. Like Central Park in New York, it is often used as a basis of comparison by managers and designers of other parks. In total, the Hyde Park complex is approximately 750 acres. These lands were once royal hunting grounds near Whitehall Palace and centered on the two tributaries of the Thames River: the Tyburn and Westbourne streams (FIG. 1). The park can be seen as a palimpsest; one can actually see the layers of design throughout the park. As one studies the history and current management of the park complex, this sensation of palimpsest grows and manifests itself in a variety of ways. At the edges of Kensington Gardens, rows of trees in several different patterns indicate a series of prescribed allées whose roots are no longer apparent and are in an odd juxtaposition with one another (FIG. 2). Green Park has a subtle and complex topographic condition conducive to casual use on one hand, while on the other tree rows overlaying this same native topography reveal connections to past royal life, or in recent times, as one of the settings for the Queen's Jubilee (FIGS. 3, 4).

Native topography and soil conditions create in Hyde Park one of the most flexible and successful open lawns of any public park. This high, well-drained ground, left seemingly ungraded, performs as an active field for soccer and rounders as office workers empty from their nearby buildings after work hours (FIG. 5). This same high ground is occasionally converted to a major event space, able to host performances such as a Rod Stewart concert during the Jubilee. The native topography of Hyde and Green parks leads to an easy and casual relationship with nearby residents and office workers, as it is performative (in the case of the high ground) and visually compelling and inviting (in the case of the concavities of Green Park) (FIG. 6).

FIG. 2: Kensington Gardens, row of trees contributing to palimpsest, 2001.

FIGS. 3, 4: Green Park, unmade terrain, 2002 (opposite,top) and London plan tree aleé, 2001 (opposite, left).
FIGS. 5, 6: Hyde Park, ad hoc recreation on the park's high ground, 2001 (opposite, right) and people using the slope down to Serpentine Lake, 2001 (above).

Made topography, however, is present in two places: along Kings Row where the Crystal Palace once stood and throughout Kensington Gardens. The former site of the Crystal Palace is framed by London plane trees, creating a spatial volume evoking the ghost of its famous former occupant. This level ground flanking the residential section of Knightsbridge is reserved for active recreation and organized sports. Kensington's topography and landscape have a different history. In a search for drier ground and air, William III moved the palace away from the Thames and Whitehall to the edge of the village of Kensington in 1688. Initial gardens formed a small complex of 17 acres. One hundred acres were added by George I in 1708, and in 1726 Charles Bridgeman began his work on Kensington, continuing through the reign of George II. With another 150 acres, Bridgeman conceived of a grand complex using flat and sloping planes of made topography to create grand vistas, walkways, and connections to Hyde Park. Complete with long allées of trees, this highly spatial yet open-ended, volumeless design survives today, but with a very different maintenance regime. In conversations with the royal gardener that I began in 1998, I learned that the staff have been perplexed over how much restoration the gardens should undergo. They did feel that its stripping of linear spatial qualities by the Victorians reduced its significance and was leading to a homogenization with Hyde Park. They also knew that, royal garden or not, their resources were limited. In a rather brilliant strategy of adaptive management, they are in the process of restoring the allées but not the fanciful zoos and gardens that occupied the interstitial landscape rooms. Here they are allowing long grasses and wildflowers to form meadows that manage human use through surface condition, increase bird habitat, and decrease maintenance cost (FIG. 7). Also, an enhanced sense of beauty and grace is evident to those of us who have followed the evolution of this park complex over the last decade.

It is interesting to observe the difference in use by park goers between this made planar topography and the found topography of Hyde and Green parks. Whereas Hyde and Green casually enfold the visitor, who can appreciate the parks from a fixed point, the user of Kensington is always in motion, as if the made topography demands a constant procession. It is only at the Round Pond, built for George I's turtles, that human agency comes to rest.

At St. James Park, the notion of palimpsest reveals itself in another way. Made grand and straight in 1660 by Andre and Gabriel Mollet at the behest of Charles I, a former tributary was aligned with Whitehall in the manner of the Grand Canal. Over time a menagerie complete with Russian pelicans occupied the southern end of St. James. To this day, even after a complete undulating remodel of St. James by John Nash in 1823 and its full introduction into the public realm, the exotic species of three hundred years ago still exist. A combination of design and management has kept one or two islands as exotic habitat sanctuaries, which along with a managed shoreline condition provide the necessary ingredients for an urban waterfowl habitat (FIG. 8).

FIG. 7: Kensington Gardens, meadow, 2002.

FIG. 8: St. James Park, remnant waterfowl habitat, 2002.

No discussion of Hyde Park is complete without addressing the Serpentine Lake. William Kent gave form in 1728 to the Serpentine as we know it today. Ostensibly done to accommodate two royal yachts, the redesign expanded the water feature significantly, uniting Hyde Park and Kensington Gardens with one of the most evocatively named water bodies in the public realm. Germane to this discussion is how the program around the Serpentine has evolved and its relationship to landscape and paving systems. The area south of the roadway bridge abuts the lower slopes of Hyde Park's high ground and has a continuous promenade, two restaurants, boat rental concessions, a swimming zone, and an exhibition zone for rollerbladers—altogether an excellent example of connected program of human-use elements (FIG. 9). A very different environment has been produced from the area north of the roadway bridge. Access to this portion of the Serpentine exists only at the main axial plane of Kensington Gardens. The rest of the shoreline is densely planted and protected from land and water access (FIG. 10). This is clearly an area for urban wildlife rather than people, at one time producing habitat and complexity of landscape in contrast to the highly programmatic and reductive landscape along the Serpentine to the south.

As a complex, these parks actually form one of the smaller areas studied in the student research seminars at the Harvard University Graduate School of Design. However, I believe that when studied closely, it is one of the more complex and successful public parks in existence today. There is no question that some of this complexity comes from being developed over the course of four hundred years, but I would contend that much of the variety and success arises from the layering and juxtaposition of the made and the unmade, or the designed and the undesigned. Although these parks have many elements in common, such as green turf, trees, and walkways, they are each different as to topography and forms (FIG. 11). Thus this amalgam joins with flexible surfaces and adaptive management to position in the public realm a landscape true to its royal roots, but through use and its publicness, growing deeper into the population's psyche with each passing moment. I believe we can imagine an England without the royal family, but not without these parks.

FIG. 9: Hyde Park, southern portion of Serpentine Lake and its linked program areas, 2002.

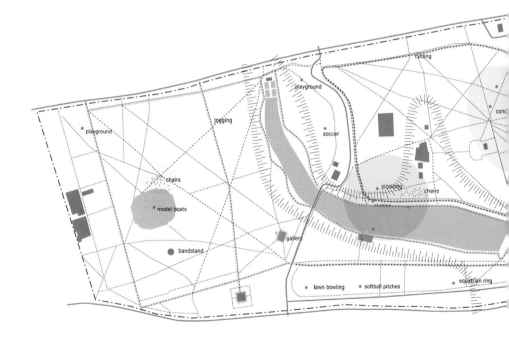

FIG. 10: Hyde Park, northern portion of Serpentine Lake with a plane from Kensington Gardens on the left and wildlife habitat on the right and in the distance, 2002 (opposite).
FIG. 11: Hyde Park complex, topography and program diagram drawn by Jason Siebenmorgen, 2003 (below).

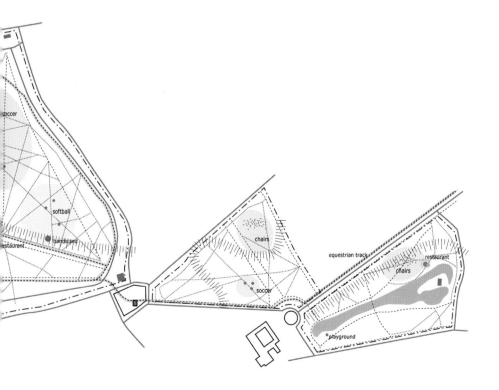

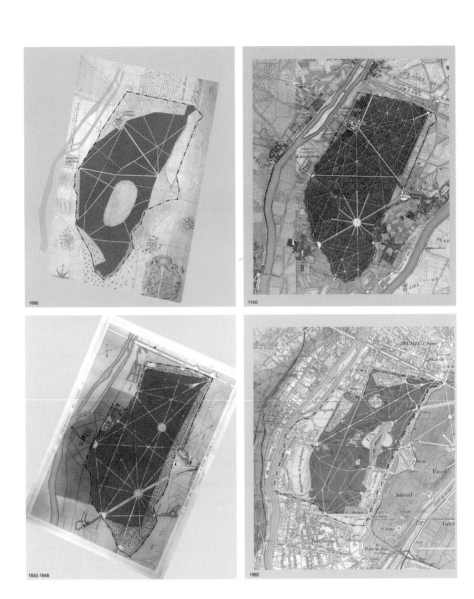

FIG. 12: Evolution of the forest (1666) to the Bois de Boulogne (1882).

The Bois de Boulogne: The Park as a Jackson Pollock

The Bois de Boulogne lies on the western edge of the city of Paris. Comprised of a former royal hunting ground and an additional 700 acres, these lands totaling approximately 2,000 acres were decreed public by Napoleon in 1852, and under the direction of Adolphe Alphand, construction began one year later (FIG. 12). Situated on the alluvial plain of the Seine, the Bois contains the remnants of the ancient Forêt de Rouvray. In what may have been a harbinger of the Bois condition today, more than a thousand years ago various communities created clearings in the forest and trails to the city. These travelways through the forest were often dangerous, and the woods were full of illicit activities. In the sixteenth century, the forest was enclosed to become royal hunting grounds and, as was the fashion of the times, straight lines were carved through the forest to facilitate the hunt and social activities of the royal parties. Alphand began by removing all but two straight travelways through the park and replacing them with sinuously curving walks and carriageways as a counterpoint to Haussmann's new axial boulevards through the city. We see this same metaphorical expression of city grid transforming to curvilinear travelways played out in Central Park fifteen years later.

As we examine Alphand's transformation of the Forêt de Rouvray into the Bois de Boulogne, one never wanders far from the forest. Even today, it is evident in Katherine Anderson's conversations with Bridgette Seere, one of two foresters assigned to the Bois, that the forest of the Bois occupies a position in the Parisian psyche that is certainly romantic, maybe biological, and bordering on the mystical (FIG. 13).[3] Perhaps borrowing from previous settlement patterns or intuitively guided by a large patch and corridor approach, the human activities or program are typically dispersed through the forest and joined by trails and paths (FIG. 14). The tradition of illicit activities in the forest continues; in fact, the police prefer that prostitution occurs in the forest or along Avenue Forêt, as it removes commercial sex from the streets.

As we travel through the forest, we also encounter horserace tracks, private tennis clubs, a tennis stadium that hosts the French Open, soccer fields, playgrounds, *boulles* courts, ornamental gardens, and restaurants. Intentionally all of these program elements are dispersed through the forest rather than linked, and even as more program demand is placed on the Bois, there is resistance to adopting a new strategy. There are few meadows, and as proudly stated by current management, this makes the Bois "unlike Hyde Park." It is also interesting to note that there is virtually no manipulation of grade. The only area where this occurs is around the Lac Superieur and Inferieur, where the few meadows appear; it is the only area where linked program exists in Alphand's original plan (FIG. 15). Probably the most inhabited and actively and casually used, the lake system presents itself as a very linear configuration with narrow islands and exotic plantings around its shallow shores. One reads and traverses the sloping grade along the shore, where the demands of flatness for water register against the cut slopes on the upper

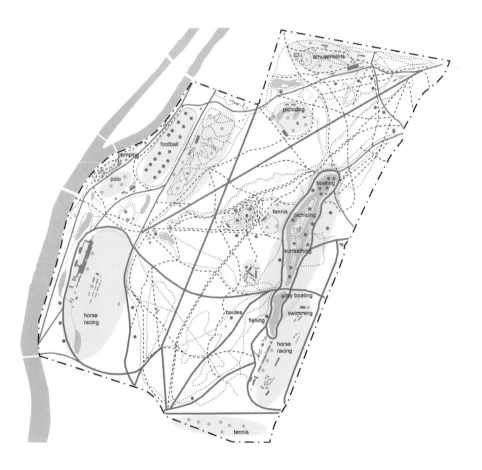

FIGS. 13, 14: Bois de Boulogne, view of the forest, 2003 (top) and diagram of topography and program drawn by Katherine Anderson, 2003 (bottom).

FIG. 15: Bois de Boulogne, sunbathers at Lac Superieur with the Eiffel Tower in the distance, 2003.

FIG. 16: Bois de Boulogne, Lac Superieur cut into native topography, 2002.

FIG. 17: Bois de Boulogne, Lac Inferieur, view of water level above native topography, 2002.

lake, producing an upper- and lower-lake promenade with intervening open grass areas used for sunbathing and picnicking (FIG. 16). The lower lake extends beyond native ground where its flatness reveals its fill condition, as Parisian boaters float on a lake above the picnic grounds (FIG. 17). Filled with boaters, picnickers, strollers, and runners, the lake area teems with Parisian life on the weekends, as it borders a dense residential neighborhood.

When measured against the rest of the Bois, this heavily used area is quite small; indeed, there is a tension between the amount of human agency taking place at the Bois and the desire to maintain the forest as a primitive zone. Current management practice is to resist further incursion of program into the park from the edges, thus keeping the forest patch more or less intact. It is probably the best strategy for now, except for the government's tendency to allow privatization of some elements within the park, thus removing them from the public realm and exacerbating a growing problem.

In 1999 an ice storm swept through Northern France, devastating many large tree plantations including the Bois. What could have been a major rethinking of the park's layout of uses in relation to quantity and location of forest was rejected after some consideration, in favor of a fairly complete replacement of the forest. Park management is using three strategies: natural regeneration or seeding as it maintains the genetic capital, dense sapling plantings in large devastated areas, and large bag and burlap plantings along major trails, paths, lakes, lawns, and other public areas (FIG. 18). Is this the right approach, or has there been a major opportunity missed that could have yielded a nuanced version of the Bois that more comfortably dealt with the demands of the users? Even though park management rather melodramatically cites people as the problem and the forest as perfect, it is the forest that is unique to this public realm. It is the forest that led to this specific site for the Bois. In the end it is difficult to fault a decision whose basic tenet is to return to the site's inherently unique quality. Although it is now less than one-tenth

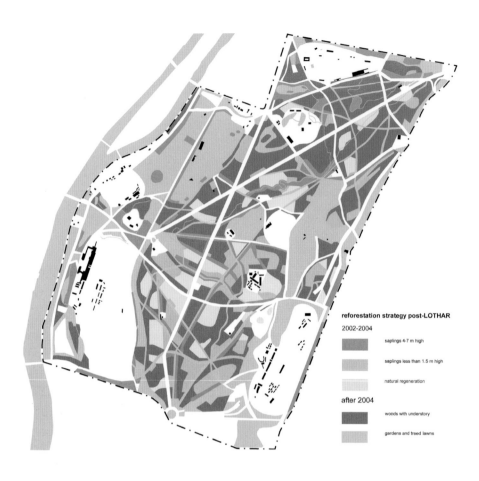

reforestation strategy post-LOTHAR

2002-2004

saplings 4-7 m high

saplings less than 1.5 m high

natural regeneration

after 2004

woods with understory

gardens and treed lawns

FIG. 18: Bois de Boulogne, diagram of 2002 reforestation strategy drawn by Katherine Anderson.

its size of a thousand years ago, if it can be protected from future encroachments and privatizations, the remnant Forêt de Rouvray could become one of Europe's sacred places.

At times, the Bois could be seen to resemble a Jackson Pollock painting, the mad slashes of roadway pulsing with commuter traffic through the park. While conjoined with splotches and drips of program to create a vivid image, there are times when the Bois is an angry canvas of combat—to wit, an early morning crash of car and motor scooter was witnessed on my second day in the Bois. Along with the curtailment of encroachment and privatization, a reduction of roadways through the Bois would ensure the survival of the Bois forest and generate easier and safer connections between the dispersed activities. Perhaps it is time to engage the mystical attachment of the public to the forest in order to save it.

Both the Bois de Boulogne and the Hyde Park complex continue to draw on the history of their sites as a source of character and identity. The next two parks addressed in this chapter, Golden Gate Park and Centennial Parklands, demonstrate a very different approach toward their sites, one of remaking and erasing identifying site characteristics.

Golden Gate Park: An Entropic Postcard

The original site condition for Golden Gate Park was approximately three-quarters sand dunes, beginning at the far western edge of the site adjacent to the Pacific Ocean, with the remaining eastern portion predominately clay loam and covered in fairly healthy native shrubs and trees. In 1870 these lands of slightly more than 1,000 acres were designated a public park by the city of San Francisco. The park was conceived at the height of the nineteenth-century park movement in the United States; the design of the largest park in the west arrived like a prototypical carpetbagger from the east. William Hammond Hall, the engineer and overseer of the park's design and construction, developed a notion of restorative pastoral scenery from the literature of the day. He also promoted the idea of a fixed picture or scene that would accommodate a predetermined social program of passive engagement with the landscape that, according to Hall, would induce the highest moral order. With these precepts of social engineering established, he set about developing the method for transforming those pesky dunes into rural picturesque scenery similar to the northeastern woodlands. By grading the dunes into a configuration adapted from the countryside of the northeastern United States and then covering them with clay from the marshes of the city, the transformation process was begun (FIG. 19).

The first problem of the forestation process was the failure of deciduous trees, which comprised a majority of the tree species. However, on the second try, conifer trees from nearby Monterey, native acacias, and a wide variety of eucalyptus imported from Australia thrived to create a unique forest of stately trees. The picture postcard envisioned by Hall arrived, but with a slightly

FIGS. 19, 20, 21: Golden Gate Park, regrading of native dunes (top); meadow and dale, 2002 (opposite); and topography and program diagram drawn by Ananda Kantner, 2003 (bottom).

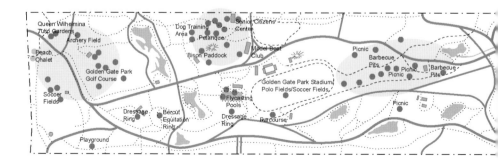

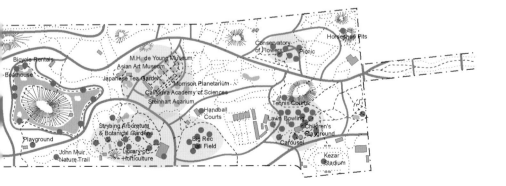

different look. The meadows were there along with the gently rolling hills and dales, but without the oak and maple forest (FIG. 20).

As to human use, Hall seemed reticent to place any of the park's limited programmatic elements in the foreground or near the city neighborhoods, and thus like the Bois we find program dispersed through the forest. Also like the Bois, this arrangement made many of the elements hard to find unless one knew where to look. Perhaps this was the intention, as Hall continued the northeastern paradigm of eschewing active recreation in favor of the meadow, hill, and dale. Americans are an active lot, and as Alexander von Hoffman noted in his historical research on the work of Olmsted and Vaux on Franklin Park, no sooner were the meadows completed than they were turned into golf courses and ball fields and the carriageways into bicycle race courses, rather than the picnicking and strolling venues as envisioned.[4] And so, slowly but surely in San Francisco a polo grounds, soccer fields, archery fields, playgrounds, a carousel, conservatory, aviary, and buffalo paddock were added, all in the manner of dispersed program in the forest (FIG. 21). In plan one can see the reduction of forest over time, to where it is at times but a scrim maintaining the illusion of Hall's original mimetic vision.

In 1894 the Midwinter International Exposition was held in the eastern end of the park. Left from this event, the park enjoys several connected program elements centered on an open-air music concourse. Here in the park's most visited area are an art museum, planetarium, aquarium, and a Japanese teahouse and gardens. This area of linked programmatic elements possesses a critical mass of activities near the city and serves as the introduction to Golden Gate Park for many users. From here, a jumping off into the rest of the park is encouraged, even though through time many of the forested areas have housed the homeless, drug users, gay cruising, illegal dumps, and roving gangs. Hall's vision of a park of passive uses encouraging moral rapture became anything but. The successful areas were quite active, though not part of the original plan, and the deciduous woodland forest did not materialize; instead, an evergreen forest provided cover for activities many citizens consider unnerving. Adding to this difficult state of affairs, by the 1970s the forest had reached the end of its lifespan, and not being a native or even naturalized condition, decay and disrepair set in (FIG. 22). Under the Golden Gate Park Forest Management Plan, thousands of trees were planted in the 1980s and 1990s. Many of the trees planted, such as Torrey pine, bishop pine, coast redwood, and coast live oak, will adapt to this environment better than some of the previous plantings. However, none of these trees exist in the sand dune formation, which is the underlying geomorphology. Add to this the resistance of nearby residents to change and the experimentation of the staff, and one wonders if in a hundred years a new reforestation effort will be needed.

Golden Gate Park is yet another example for the designer and other parties involved in the creation and stewardship of large parks that active program elements cannot be eschewed; they will come in a public realm that

FIG. 22: Golden Gate Park, decaying cypress and eucalyptus with replanting in the foreground, 2003.

resides in democracy. It is also a prime example of the difficulty in creating a sustainable landscape when the site's base geomorphology is denied. Although it would be controversial, I would suggest a close look at returning at least part of the western portion of the park to dunes, or Golden Gate Park will be forever condemned to an unsustainable future and never outgrow its title, an entropic postcard. Or is it too late to recognize the folly of denying a site's basic biological characteristics?

Centennial Parklands: The Park in Constant Flux

In a contrasting approach, Sydney's Centennial Parklands are returning a small portion of the park back into swamplands. Dedicated in 1888 to celebrate Australia's one hundredth anniversary, Centennial Parklands became the center of a 900-acre park complex that includes Moore Park to the east and Queens Park to the west. Centennial Park transformed Lachlan Swamp and its surrounding sandstone bluffs to an Australian vision of the nineteenth-century public park. The park has unfinished qualities, and one could argue that a public's love could render it as unsustainable as Golden Gate Park.

A central ring provides the foci of Centennial Park, perhaps to recall the riding ring of Hyde Park. Surrounded with Moreton figs, this circuit was meant for strolling and carriage rides and encircled large fields for calming the eye as well as the spirit (FIG. 23). In what has become familiar in large parks of the new world, things began to change right away. Many of the exotic Victorian plantings withered and died in the Australian heat, even as the swamp and bluffs were destroyed to create what was thought to be the proper environment for Sydney's first public park.

Not long after the initial plantings died, the public demanded that the grounds be opened up for active sports such as cricket and football. Charles Moore, as director of the Royal Botanical Garden, undertook the first major change in Centennial by adapting its edges to active uses, primarily sports, while maintaining a central open area of flexible green. The next director, Joseph Maiden, while keeping the ring of figs, created the water and plant system many Sydneysiders enjoy today. Resurrecting the water draining into the site as a series of linked lakes and establishing extensive plantations of native trees, Maiden began to recognize the natural forces prevalent on the site (FIG. 24). He even established shade breaks to shelter users from the oppressive summer heat between the large fields. After Maiden's death in 1925, the park lost its keeper. As the government let Centennial Park degrade, it was the public users of the park who rallied to its support. Today as one circulates through the park, there is the ever-present ring of figs with an equestrian ring, jogging and walking ring, and car and bicycle way. Continually used by a very active populace, the ring encircles an alternating landscape of open fields and water catchments teeming with birdlife. In fact, the ring of figs became such an iconographic piece of the city's public realm that when Bicentennial Park was created in the western suburbs in 1988, its central feature was a ring

FIGS. 23, 24: Centennial Parklands, ring (to the right) and replanting of native trees, 2002 (top) and Joseph Maiden's reintroduction of water system and native plantings, 2002 (bottom).

FIG. 25: Centennial Parklands, reintroduction of a stand of native eucalyptus trees, 2002.

roadway lined with fig trees, and the Olympic Plaza for the 2000 Olympics was also lined with fig trees.

When the government sought to convert much of Centennial to an Olympic training facility in the 1970s, the public demanded separation of Centennial Parklands from the government, and Centennial Parklands Trust was created in 1983. Headed by Peter Duncan, the Trust recently commissioned a heritage plan that is controversial, as the politics surrounding Centennial are contentious at best and viewed with suspicion by nearby residents and users. In another of those only-in-the-new-world situations, you have a park that has been continually changing from its very beginning and is changing now with the reinsertion of bush and swamp, yet with its heritage status, what is historical and therefore impervious to change or adaptive management may stifle what has heretofore been a fairly flexible ecological and programmatic environment (FIG. 25). The heritage plan is struggling with the relationship between ideas that would increase the park's sustainable characteristics but change the nature and appearance of some areas. If those in charge of the Serpentine and Kensington Gardens can engage in adaptive management, one wonders why Centennial Parklands may be frozen in some version of someone's proper time. Is it possible that too much public input leads to freezing a large park in moments that may be inappropriate? It seems feasible to retain the vibrant qualities of the ring and the park's edges while reexamining the intertwined relationship between swamp, stormwater runoff into linked catchments, and bush.

Amsterdam Bos: The Park as Diagram

Amsterdam Bos is a 2,212-acre park made from agricultural lands that were once sea, and previous to that, peat bog. Conceived as parklands in the 1920s and built with manual labor during the 1930s and completed, if we can use that term, in the 1950s, the Bos in its fourth version of a landscape is a modern diagram that incorporates sports programming, casual use, a sustainable

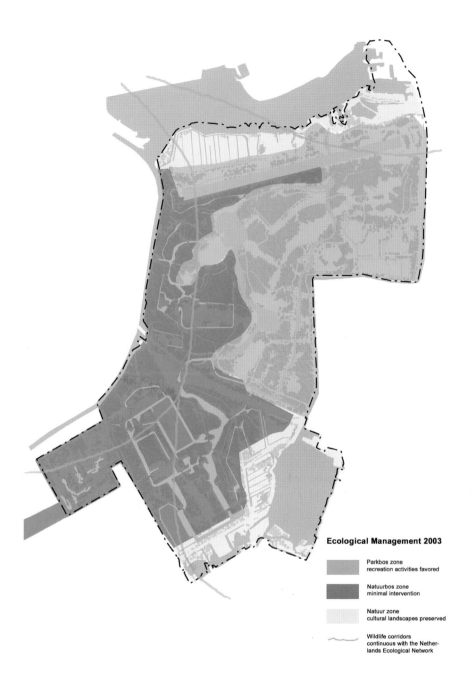

Ecological Management 2003

Parkbos zone
recreation activities favored

Natuurbos zone
minimal intervention

Natuur zone
cultural landscapes preserved

Wildlife corridors
continuous with the Netherlands Ecological Network

FIG. 26: Bos Park, diagram of ecological management zones drawn by Rebekah Sturges, 2003.

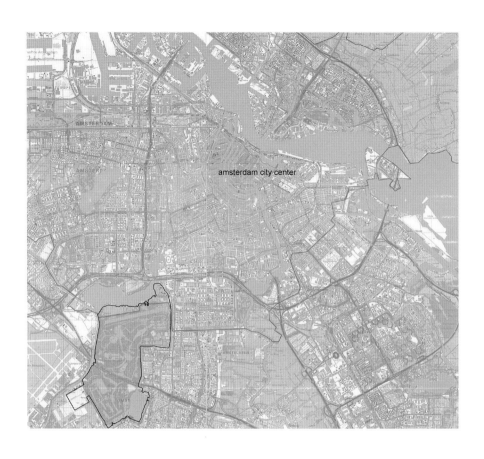

FIG. 27: Bos Park, map of the region depicting "lungs of the city" strategy.

FIG. 28: Bos Park, topography, water, and program diagram drawn by Rebekah Sturges, 2003.

forest, and remnant agricultural lands and waterways. Overlaying the park as diagram are the concepts of "park bos," where the designed-for-recreation landscape is apparent, and "nature bos," where the forest is dominant and with the agricultural fields and waterways provide rich habitat (FIG. 26). As a regional plan, the Bos was part of a larger strategy of green spokes ameliorating the urbanized environment—another version of the green-lung strategy (FIG. 27). As a design, the Bos plan by Cornelis Van Eesteren and Jacopa Mulder combines the notions of the *volkspark*, a functionalist park paradigm from Germany that embraces active sports and recreation, and a modern pictorial landscape. In so doing, it introduces a program strategy that straddles the linked and dispersed methods of placing human activities in parks (FIG. 28).

Two other things distinguish Amsterdam Bos. First is an initial establishment of more than 1,500 acres of forest that seeks to be sustainable at maturity. Second is an embracing of a previously made landscape. It is as if the latter was treated as a found object. The original plan envisioned the entire park as what is now "park bos." But due to a fortunate economy of means, or phasing, or recognition that the entire landscape did not need to be remade—it is unclear which—the southern portion of the Bos has a quite different character (FIGS. 29, 30). The built park has two distinct matrices. The simple definition of a matrix is that of a mold in which a thing is cast or shaped. I have come to believe that the ability to have within one design, particularly at the larger design scales, sometimes disparate forms and site conditions is critical to long-lived complexity. The Hyde Park complex is an excellent example of a subtle version of this effect. Amsterdam Bos is a more dramatic version. In other words, a large park does not have to follow a singular system. I would go farther and state that a transition is not necessary between two or more sets of systems. Indeed, a juncture can often produce heightened diversity, whether it be visual or biological.

Little has been said in the limited amount of critical literature on this park regarding the presence of matrices within the park and what it means. However, much has been made of the forestation plan, the operations that created it, and its open-ended nature. But I wonder if it isn't the presence of two matrices overlain with a management plan that straddles the two (rather than reconciling them) that instills the park with its experiential qualities, as much if not more than signified operations and forest.

The initial plantings of the forest consisted of fast-growing trees such as poplars and alders in a grid to nurse the slower-growing hardwoods of beech, oak, ash, and maple. Due to the careful coordination of drainage, soil types, and tree species selection, when one wanders from the prescribed circulation paths and into the forest, the mature forest is striking in its health, complexity, maturity, and regeneration (FIG. 31). Here designers and scientists have intelligently set in motion natural processes that will serve the public well as they enjoy the forest for its foliage, habitat qualities, and fabric in which the active and casual use elements are embedded.

FIGS. 29, 30: Bos Park, 2002, remnant polders (top) and remnant agricultural fields still in operation (bottom).

FIG. 31: Bos Park, sustainable forest at approximately fifty years, 2002.

It is important to distinguish between process as something that is ongoing, as in the bos forest, and an operation or linked series of operations. Process is an ongoing and always changing state, while operations are a set of performances derived from the act(s) of making and are not open-ended. Amsterdam Bos presents one of the most readable operations where one clearly senses the relationship between the large rowing basin, a borrow pit, and the hill (FIGS. 32, 33). Given the dramatic flatness of Holland, a hill of any dimension is striking. However, this is not an open-ended process where the designers have let go of authorship, at least as far as this hill is concerned. It certainly is in the correct location, as it provides the counterpoint to the panoramic green and boathouse, the organizing element of an extensive bicycle and pedestrian circulation system, and the central viewing platform from which over half of the park can be seen. It is shaped to form a cone (and is not located adjacent to its borrow source). Anita Berrizbeita has extolled this modern park as an exemplar of process in her brilliant but at times overreaching essay, "The Amsterdam Bos: The Modern Park and the Construction of Collective Experience":

> [T]he system of production prevails as the basis for design and for achieving signification, and the designer's role is transformed into one in which they set in motion the processes that will complete the construction of the park...Van Eesteren and Mulder relinquished control of the final form to processes of succession and hydrology.[5]

In my estimation, however, Van Eesteren and Mulder hardly relinquished control. As a designer and builder, I look at the hill as an iconographic element constituting a major design feature of the park. It was intentionally formed and placed. Even its surroundings were determined. I certainly agree that the exact tree species immediately adjacent to the hill were not precisely chosen by the designers, but this is a design-formed landscape that exposes its operations and harnesses the process of sustainable forestry (FIG. 34).

FIGS. 32, 33: Bos Park, 2002, rowing basin, source of fill for the hill (top) and water and meadow with slope up to the hill in the background (bottom).

FIG. 34: Bos Park, hill with forest framing space, 2002.

I believe this to be a design method laden with content and full of an exacting intent with regard to location, spatial volume, and surface character. It is not an example of "look ma—no hands," to quote design theorist Robert Somol's turn of a familiar phrase, setting in motion an open-ended process where the final, evolving form is undefined.[6]

Operations are distinctly different from open-ended processes, as I believe the example of the hill shows, but as we study the forest, which is an ongoing process, it exists as a fabric or a background upon which elements of program or space occupy the foreground. So rather than look at Amsterdam Bos as a produced landscape with undetermined form, as a designer and builder I see a landscape of complexity through the coexistence of differing states of made. So not only does the Bos have two distinct matrices, it also uses two distinct design strategies: one, a set of operations guided by design as sure as any exact placement of elements and spaces can be; and two, a sustainable forest that may ebb and flow, but that provides a constant background for human activities in this public realm.

Parc du Sausset: The Park as Narrative

The Parc du Sausset lies well outside the city of Paris, between Aulnay-sous-Bois to the south, a community dominated by public housing and populated by immigrants from the former colonies, and Villepinte to the north, a middle-class community populated primarily by the French. This approximately 500-acre park is suburban and was the winning submittal of Michel and Claire Courajoud. The goals of the competition were twofold: to preserve a large tract of open land from rapidly encroaching development and to reinstate the region's landscape identity. Construction began on the park in 1981.

Surprisingly, the Parc du Sausset has received little critical attention, to the point that most of us are familiar with only a handful of images: the water towers, the forest from the air, the train moving through the park, and maybe the wetland. Whether due to phasing or lack of initial photogenic qualities, there is an entire piece of the park that is relatively unknown—the *bocage*. The Parc du Sausset is comprised of three parts expressed by different matrices: the urban park, the forest, and the bocage (FIG. 35). The notion of the park as a narrative comes from these clear typologies.

The urban park lies at the entrance from Aulnay-sous-Bois and connects the village to the train station with a simple linear pedestrian way. Highly diagrammatic, the urban park relies on a 100-meter grid as its major structuring device. To the designers' credit, these landscape rooms are used as they were intended—for active recreation and picnicking (FIG. 36). At the south and east of these rooms lies the wetland, which receives all the runoff from the urban park. To the north there is the forest. An implied connection of land bridges points to the forest, but they are only implied. Access to the forest is across a fairly busy highway bisecting the park and inhibiting what could have been an easy and casual relationship between urban park and forest.

FOREST

BOCAGE

URBAN PARK

FIGS. 35, 36, 37: Parc du Sausset, planting diagram with park typologies drawn by Anna Horner (top); 100m x 100m room in urban park (above, left); and forest blocks, 2002 (above, right).

FIGS. 38, 39: Parc du Sausset, 2002, monoculture (top) and towers (bottom).

The forest is essentially uninhabited. It stands as geometric blocks of monoculture in what seems to be a clear expression of the Corajouds' preoccupations, as voiced by Michel: "What interests me is not what nature has to offer in terms of spontaneity, but rather worked nature, guided by human intervention (FIG. 37)."[7] The forest and its meadows are indeed beautiful to behold, and they may well be French typologies, but it is difficult to peel back some of the Courajoud rhetoric regarding flexibility, time, and human intervention, and reconcile that rhetoric with a fairly intensive maintenance regime that is readily apparent throughout the forest.[8] The intent seems to be the clearing of all understory, and yet over time the forest managers have suggested that invasion of one monoculture or another may occur—concepts that appear to be contradictory (FIG. 38). At the forest's edge stand the water towers, familiar icons of the park and anchoring devices to the matrices of Parc du Sausset (FIG. 39). An iconic reference on the skyline, they stand as reminders that no matter how entranced we become with landscape methods, dramatic vertical elements engage the viewer in ways that are powerful and at times ubiquitous.

Returning to the urban park, the wetland receives and somewhat cleanses the stormwater runoff from the park (FIG. 40). Connected to the wetland is a lake separating a large green sloping surface from an ingenious earthwork that links the urban park and the bocage and forest by connecting to a small bridge over the railway. Using borrow from the lake excavation, this landform-cum-landbridge rises up to the highest point in the park and creates sweeping vistas (FIG. 41). Similar to our conical hill at Amsterdam Bos, this arced landform is placed and shaped in such a way to perform tasks. Neither landform is the result of an open-ended process, but a design resulting from a series of controlled operations.

The bocage may be a different story, as it possesses a limited amount of open-ended design at the margins of its agricultural operation (FIG. 42). A lightly transformed agricultural zone where the introduction of hedgerows and active farming provides a glimpse of traditional French agrarian practices and landscape types, the bocage, unlike its counterparts, the forest and the urban park, leaves the remaining meadows largely to their own devices. Wildflowers, shrubs, and trees indicate a landscape patterning of natural seeding, providing a welcome relief to the geometrization of almost all other vegetal plantations within the park (FIG. 43).

Parc du Sausset's total is much greater than the sum of its three parts. As a landscape experience, it is cinematic and educational. Whether nearby residents value the regional typologies as much as they value the open surfaces for extended family picnicking and active recreation is questionable. Must the park user decode the regional typologies to value the landscape? I think not, but with onsite education available, the opportunity for parkgoers to learn of the landscape's deeper meaning at Parc du Sausset is a plus. From a designer's viewpoint, the imperative to produce several of the region's

FIGS. 40, 41: Parc du Sausset, 2002, constructed wetland that partially cleanses water before it flows into the adjacent lake (top); and lake, footbridge, and landbridge (background), 2002 (bottom).

FIG. 42: Parc du Sausset, the bocage at its entry, 2002.

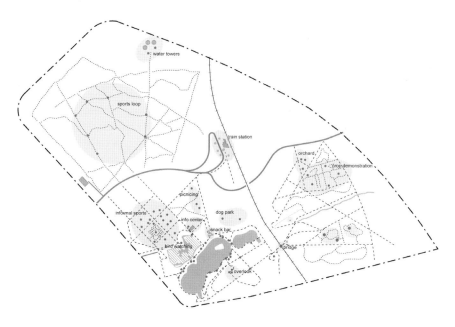

FIGS. 43, 44: Parc du Sausset, meadow in the bocage, 2002 (top) and park matrices and program diagram drawn by Anna Horner, 2003–2006 (bottom).

typologies is another method to realize the complexity that matrices can bring to a large park. Whether the high degree of maintenance and management can be sustained is a question whose answer may depend on the educational objectives. At any rate, as a designed-form landscape the linked operations of wetland, lake, earthwork, and connection underlie the three matrices of Parc du Sausset in a beguiling way that we are only beginning to understand (FIG. 44).

Landschaftspark Duisburg-Nord: The Park as Artifact and Artifice

Landschaftspark Duisburg-Nord has received a good deal of critical attention since its opening. On my first visit, two tables of landscape architects were at the outdoor restaurant and, including ourselves, outnumbered all other visitors. Opened in 1994, Duisburg was the result of a competition sponsored by the IBA in 1991. Designed by Latz + Partner, the 550-acre site was formerly the property of Thyssen, which produced steel there for approximately a hundred years. Set in context, Duisburg is meant to be an exemplar of the IBA Emscher Park, an abandoned industrial zone along the Emscher River, which the IBA wished to revitalize through green ecological frameworks surrounding the industrial monuments of the region. Hence we have the now familiar images of the artifact, together with the Latz water-recovery system—the artifice—to give us the park as artifact and artifice (FIG. 45). Many of the issues embedded in this project lie with the goals of the IBA. The focus is on repositioning the industrial past of the Emscher Park through adaptive reuse and with a limited amount of resources given post-reunification economics. As a result of the park's limited resources and the largeness of the site, there is a seeming inability to transform the post-industrial artifact and the necessity of using a laissez-faire approach to remediation once away from the project core. One begins to realize that although there is a lot of language concerning ecology and concepts of greening, there is very little remediation. Landschaftspark Duisburg-Nord uses capping to cover a few of its most toxic areas and an attenuating landscape strategy over the rest. Whether this will take a hundred years to remediate or thousands is anyone's guess.

The park may be thought of as three distinct zones. Two zones respond to the industrial past. One, the industrial machinery, familiar to us through images in publications, has been opened to the public for access, interpretation, and adaptive reuse such as diving in the gasometer and wall-climbing in the ore bunkers. Two, the wilderness area, lesser known, features a potpourri of plants from around the world that arrived as seeds via rail cars over the life of industrial use on site (FIG. 46). The third zone is derived from the immediate context where gardens, playgrounds, a community center, and recreation fields abut a residential neighborhood.

Among the ore-processing machinery, one finds several areas of cultivated landscapes, tree bosques, hedges, and wavy parterres. A distinct lack of maintenance creates a blurring between the cultivated zones and what appear to be volunteers in and around the machinery (FIG. 47). As you move around

FIG. 45: Landschaftspark Duisburg-Nord, water treatment in foreground (artifice) and designed landscape with industrial artifact, 2002.

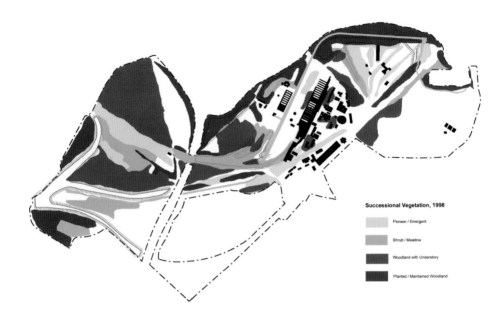

Successional Vegetation, 1998

- Pioneer / Emergent
- Shrub / Meadow
- Woodland with Understory
- Planted / Maintained Woodland

FIG. 46: Landschaftspark Duisburg-Nord, diagram of vegetation types drawn by Gina Ford, 2003.

the periphery of machinery, water-collection zones begin to appear, often as fairly romantic water gardens complete with iris and lily pads. The water is ameliorated to some degree and circulates to a long linear canal that was once used for sewage but under the Latz scheme is diverted away from the canal. On a particularly hot day this summer, local youth found refuge from the heat and turned the canal into a swimming pool. Near the canal a pleasant exhibit explains the water system, but when juxtaposed with the nearby coal-storage bins and slag heap, the artifice seems clever but trite (FIG. 48). More disturbing is the isolation of interpretation to the industrial process and stormwater collection system. There is no mention of the Jewish slave laborers held here against their will to further the manufacture of steel for the Nazi war machine, many of whom perished in the process. The story of Duisburg is incomplete without its full cultural context, and is also perhaps a troubling celebration of the industrial sublime.

Moving to the wilderness area, it is typically unpopulated except for the occasional bicyclist or landscape architect. Somewhat impenetrable, it is a landscape to move through—which is probably a good thing, as it covers some toxic areas, which I believe was Latz's intention. Above the wilderness and connected to the machinery are the railroad tracks that form a circulatory spine through the park. This adaptive reuse is extremely successful, as it provides an elevated procession with recognition of the former industrial process to form a major park element (FIG. 49). Its transformation is a potent signifier of past and present. The designer's intent was that it also connects to future phases across the highway. However, access via the railway is blocked, and one must return to the car and hunt through the neighborhoods to find the rest of the park. It seems that the authorities will develop the portion meant for fields and meadows into a commercial development centered on an IKEA store.

As one peels back the layers of the sublime imagery of Duisburg, its history is not as benign as the park would lead one to believe, but the idea of keeping an industrial artifact as a monument open to the public has a certain resonance with our culture as the face of industry changes. Design interventions, however, can easily seem trite next to the brute strength of these machines.[9] One wonders if interventions should be as close as possible to the scale and power of the industrial process once present on the site, and whether they should be tied to remediation wherever and whenever possible.

As true to site condition as these projects may be, at what point does Emscher Park—or for that matter, say, the Pittsburgh environs—have enough monuments to the industrial past? At what point are we as a society better off with a removal and remediation strategy so that we can embrace a future using land in as smart a way as possible?

Conclusion

Sites are the fundamental building blocks of any landscape. In the case of large parks, large sites often determine physical characteristics and to some degree

FIG. 47: Landschaftspark Duisburg-Nord, designed landscape and industrial artifact, 2002.

FIG. 48: Landschaftspark Duisburg-Nord, brute power of post-industrial landscape, 2002.

FIG. 49: Landschaftspark Duisburg-Nord, attenuated wilderness, 2002.

limit the ability to successfully create radical change. Many projects from past centuries began with sites that had special, sometimes complex characteristics, which gave them their iconic qualities. The natural topography of the Hyde Park complex or the forest at the Bois de Boulogne or the meadows and drumlins of Prospect Park are prime examples. Not much more is needed to establish a great place in the hearts and minds of the public. To be certain, there are issues of relation to city, gateways and entrances, circulation, and program type and location and their relation to a sustainable landscape, to name a few. But these should all be looked at with an emphasis on keeping the special qualities of the site intact, as at the Bois, or in the case of Prospect, extending the qualities of postglacial terrain toward the public realm. Golden Gate and Centennial parks would be healthier, more sustainable, and unique to their regions if the original designs started with the extant site rather than ideas from other places and times.

In the twenty-first century, however, it will be rare for a designer or design team to come across the great park sites of yesteryear. Today we are making parks from landscapes that range from the artificial (such as piers) to landscapes from blighted industrial areas (such as railway yards and waterfront parking lots and warehouses) to extremely toxic landscapes (such as Superfund brownfields and nuclear waste sites). These sites will require tremendous amounts of transformation if we intend them to truly enter the public realm, particularly if they lack the monumental if troubled qualities of Duisburg or the subtle subliminal qualities of Montréal's waterfront that inspired an archeological design strategy by Peter Rose. Many of these sites will be nothing more than mundane urban detritus. At issue is the formation of the site's character, its program of human activities, and its sustainability characteristics, particularly in regard to maintenance regimes. These subsets of the public park are inexorably entwined, and Amsterdam Bos is an extraordinary example of success in this regard.

As important as program is to architecture, it is a neglected stepchild in landscape architecture. In academia it is often perfunctory at best, left unexplained and unexplored by many faculty; it is no wonder the student ignores it or treats it without inspiration in the studio. In practice it is often a menu that practitioners use to appease a public dominated by special interests and a "me-first" mentality, as well as disguise a lack of purpose, leadership, and vision. Their plans often have the look of a seven-course meal purchased from a cafeteria. I challenge the profession to investigate the possibilities of human measures, events, and adjacencies in a far more rigorous and creative manner than we have throughout our profession's history. There is more here than laying out fields or plunking down amphitheaters. We need to explore ideas of multiplicity, flexibility, temporality, and unforeseen agencies, as well as program's relationship to place.

The relationship between sustainability and natural processes can be murky. On one hand, there are vibrant examples, such as the forestation of

FIG. 50: Parc du Sausset, 2002.

the Bos, and on the other there are projects where the rationale of "just let it go" prevails, such as the impenetrable wilderness area at Duisburg. I believe that an open-ended process that harnesses natural systems in a meaningful way for the overall development of a park, such as stormwater cleansing or forestation, is a great thing, but it is not the sine qua non. To the designer who believes the letting go of their version of process is sufficient to be the subject or character of a park in the public realm, I submit that this notion is narcissistic. The means are important, but only as to the end they lead to. I have not abandoned process, but rather learned where it belongs through the act of building.

Embodied in the development of a site's character are notions of place, multiplicity of form or matrices, and spatial volume and surface character. Swirling around in this milieu are of course program and process. I think the exposure of operations—in particular, linked operations—provides a working method full of content and resonant with tectonics. This will allow the designer or design team to move beyond the empty gestures of formalism and the tyrannies of ubiquitous process. And while I may rail against "letting it go" as a dominant strategy, it is okay to leave parts of a site or project unmade or undesigned, as evidenced by the topography at Hyde, the agricultural fields and waterways at Bos, and the bocage at Sausset (FIG. 50). As a designer who builds and an academic engaged in research with great students, and after documenting most of these parks in person, I have come to believe that our slavery to oneness, the unified plan, the need to remake an entire site, and abhorrence of program keeps us from realizing the full complexity and diversity that parks can have. All of these issues are exacerbated by the sites that we have for parks in the post-industrial landscape. Sites that have been made and remade, such as our abandoned waterfronts and railyards, have little of the character that can resist the unified plan or diagram. Reifying an industrial past and premiating a remediation system lack the potency to fulfill a range of human activities and/or wildlife habitats that great large parks possess. These sites will require an independent understanding that resonates with the sites' physical histories and contexts. We must move beyond our self-induced limitations toward an intertwined agenda—an agenda that celebrates human activities, creates and explores sustainable landscapes, and develops complex matrices of form and type, all in an effort to engender within the public realm those great places that capture the minds and hearts of humanity and propel a public park forward for centuries.

Epilogue

In 2005, I had the opportunity to test the various ideas presented in this chapter. Orange County, California, held an international competition to submit designs for an 1,100-acre park on a portion of the abandoned El Toro Marine Air Base.[10] Hargreaves Associates, with Morphosis and Arup, was one of seven teams shortlisted.

The former air base is on an extraordinarily flat site. Runways and hangars are still in place, and all water has been diverted away from the site. The scheme used a linked series of operations to transform the site into parklands while taking advantage of many aspects of the existing site conditions (FIGS. 32–36). Stream corridors were proposed to bring water back into the site. The excavated soil was dispersed on the site, with a datum landform set against the native 2 percent slope. Water, the lifeblood of our scheme, was proposed to create native riparian corridors suitable for wildlife and was captured and stored in linear basins, using solar-powered pumps where runways once existed. Plant nurseries for use during the parklands' decades-long development were proposed in a similar fashion. Other runways became parking lots and linked programmatic areas ranging from organized sports to casual rollerblading, model airplane flying, etc. Two of the major hangars were kept to house cultural institutions, and an infrastructure of bridge and pedestrian ways was placed in the scheme to link public transportation to the cultural institutions and then on to the parklands.

The most perplexing problem was what kind of landscape could, or should, exist on the tablelike plane of site above the riparian corridor. The native community of grasslands and oak trees is a fragile one. Western meadow grasses are spiky and not particularly people-friendly. They can also be destroyed by even moderate human activities. We developed the notion of irrigated patches within the meadows and oak trees, whose exact size and placement would be determined by the demand for cultural and recreational activities over the fullness of time. We strove to pursue flexible strategies for the majority of the park's development over time, as the amount of active programming, nature, and cultural institutions was unknown. Time would allow us to understand demand and draw on additional funding opportunities. However, certain elements that could be linked through operations and their form were by and large determined. Our hope was to develop a strategy that could involve more nature, activities, and culture in later plans, either singularly or in various combinations.

No, we did not win, but having the chance to envision site, linked operations, and adaptive management of natural resources, activities, and cultural institutions into the future was invaluable. We never "let it go"; it has all the design intent of form, surface, and materials. It is, however, a dynamic master plan with flexible trajectories over time.

NOTES

1. I would like to thank Anna Horner for her assistance in converting my Large Parks lecture into text. Beginning in spring 2002 and continuing through fall 2004, a research seminar was formed at the Harvard University Graduate School of Design to create case studies of large parks, both historic and contemporary and within reach of the funding available to us by the Penny White Fund. Working with Julia Czerniak and a group of students—Katherine Anderson (Bois de Boulogne, Paris, France), Gina Ford (Landschaftspark Duisburg-Nord, Duisburg, Germany), Ananda Kantner (Golden Gate Park, San Francisco, CA), Anna Kaufmann Horner (Parc du Sausset, Villepinte, France), Emma Schiffman (Centennial Parklands, Sydney, Australia), Darren Sears (Parque Nacional da Tijuca, Rio de Janeiro, Brazil), Jason Siebenmorgen (Hyde Park complex, London, England), Rebekah Sturges (Amsterdam Bos, Amsterdam, The Netherlands), Caroline Chen (Tiergarten, Berlin, Germany), Lara Rose (Casa de Campo, Madrid, Spain), and Michael Sweeney (Chain of Lakes, Minneapolis, MN)—we began the process of selecting and researching parks. We set out to find a few lesser-known parks, as well some that have already been analyzed but perhaps their seminal quality had not yet been recognized. As a result, some obvious large parks such as Olmsted and Vaux's Central Park and Prospect Park were not selected, given the extensive body of existing scholarly research. The students each researched and visited their selected park; their research informs this paper and was the basis of an exhibition at the GSD in spring 2003. I was able to personally visit and photograph seven of the parks during the summer of 2002: Hyde Park, the Bois de Boulogne, Golden Gate Park, Centennial Parklands, Amsterdam Bos, Parc du Sausset, and Landschaftspark Duisburg-Nord.

2. Here, I consider made places to be the products of cultures or designers, as opposed to unmade places, which are wild remnants of natural systems. Designed places are characterized by intentional forms, while undesigned places could be the byproduct of other processes, such as industrial operations.

3. Katherine Anderson, a participant in the Large Parks research seminar at the Harvard University Graduate School of Design, interviewed Bridgette Seere as part of her research on the Bois de Boulogne.

4. Alexander von Hoffman, "'Of Greater Lasting Consequence': Frederick Law Olmsted and the Fate of Franklin Park, Boston," *Journal of the Society of Architectural Historians* 47 (December 1988).

5. Anita Berrizbeitia, "The Amsterdam Bos: The Modern Public Park and the Construction of Collective Experience," in James Corner, ed., *Recovering Landscape: Essays in Contemporary Landscape Architecture* (New York: Princeton Architectural Press, 1999). On a personal note, although I am critical of certain aspects of Anita's essay, I am grateful to her for bringing attention to the Bos Park. Her essay inspired me to visit the Bos and to begin thinking and writing about large parks and their development.

6. Robert E. Somol, "All Systems GO: The Terminal Nature of Contemporary Urbanism," in Julia Czerniak, ed., *Downsview Park Toronto* (Munich and Cambridge, MA: Prestel and the Harvard University Graduate School of Design, 2001), 134.

7. Michel Corajoud, "Parc de Gerland Lyon," *Le Moniteur Architecture* 116 (May 2001), 116, 77.

8. *Management* means monitoring design strategies over time and may include sequencing of operations and program development; *maintenance* means keeping a landscape in an exact or static condition.

9. See Linda Pollak's essay, "Matrix Landscape: Construction of Identity in the Large Park," on sublimity in parks, in this volume.

10. The rest of the site is currently being developed.

FIG. 1: Bois de Boulogne, Paris, illustrative plan drawn by Katherine Anderson, 2003.

RE-PLACING PROCESS

Anita Berrizbeitia

Large urban parks are complex and diverse systems that respond to processes of economic growth and decay, to their own evolving ecology, to shifts in demographics and social practices, and to changes in aesthetic sensibilities. Because of their size (defined here as having at least 500 acres in area) their location (often close to dense urban environments), and their site histories (such as former industrial zones that need remediation to make them suitable for recreation), these parks require a process-driven design approach that does not intend to provide a definitive plan for the site as much as it seeks to guide its transformation into a public recreational space. Because the design and construction of large parks take years, if not decades—often with changes in public administration and funding in the interim, and lengthy public processes that require ongoing revisions—designs are open-ended, incorporating diverse approaches and uneven levels of intervention and management. They focus on frameworks that adapt to changing conditions rather than forms composed to conform to an aesthetic whole.

Yet for all their susceptibility to the ebb and flow of urban circumstances, large parks remain fundamental to cities, not only because they take on infrastructural and ecological functions displaced from densely built centers but because they are distinct, memorable places. They absorb the identity of the city as much as they project one, becoming socially and culturally recognizable places that are unique and irreproducible. Those large public parks that we are continuously drawn to as designers, ones that have captured the imagination of writers, artists, social historians, and philosophers, and that continue to be used intensely centuries after their making, have in common seemingly contradictory characteristics: they are flexible, adaptive, socially dynamic, emerging sites, and they are also visually powerful, unforgettable places. They are the product of deliberate decisions that leave them open-ended in terms of management, program, and use, and they result from equally conscious decisions that isolate, distill, and capture for the long term that which makes them unique. This chapter examines the relationship between process and place. More specifically, it explores how process-based practices, those that leave the site open to contingency and change—a contemporary requisite of large and complex sites—also incorporate strategies that accentuate a place's enduring qualities.[1]

On Place and Process in Landscape Architecture

Landscape architects have traditionally understood place through aesthetic frameworks that explain its distinguishing physical qualities and character. This approach to place has been fundamentally singular and static; it emphasizes landscapes as purely visual and, furthermore, typically deals with the site during the present (the time of the project), implying that places are little more than scenes frozen in time. This notion of place was challenged during the last quarter-century, when landscape architects expanded their scope of interests and methodologies, acquired a self-consciousness about the discipline's own language that had been unprecedented in the twentieth century, and, through ecology, became keenly aware of the dynamic complexity of the medium. Whereas for contemporary landscape architects the larger cultural zeitgeist called for an anti-essentialist and pluralist notion of landscapes, place had retained its connotations of singularity and presence, received primarily from regionalism and preservation, and, as subject of inquiry in design, it was put aside as too binding to foster creative thinking. More specifically, landscape architects drew from recent development in geography and art to expand their interpretive tools.

Starting in the 1980s, geographers generated a wide array of frameworks through which to interpret place.[2] For humanist geographers, place is fundamentally tied to questions of human experience, of cultural meaning—an object for a subject, something to behold, visually and emotionally. For cultural geographers, place acquires meaning through the events and social practices that occur within it: shared experience is as important, if not more so, than physical attributes. Cultural geography also contributed the notion of place as contested ground, where, as James Duncan and David Ley write, "a constellation of economic interests, power relations, cultural dispositions and social differentiation... constitute the character of place."[3] For social geographers, place is the result of a dialectical relationship between individuals and physical space, in which individuals or institutions shape places that in turn shape them through the social practices that occur within them.

Site specificity, although like place bound to an actual location, also became a fruitful model for addressing issues of the site and its perception in innovative ways. Unlike place, however, site specificity did not seem to claim to be all-inclusive, historically bound, or totalizing about a place. It only engaged one, or sometimes a few, aspects of it, such as scale, topography, the earth's position, or the ephemerality of color and light. In other words, it did not claim to be a definitive portrayal of a location, as place seemed to imply, only the artist's conception of that site at a particular time. It was essentially arbitrary in terms of choice of subject, or means to reveal that subject. It was never intended to blend in, to be conciliatory toward what was there. From that point of view, site could be anything, although place was one thing: the existing visual character of a site. Site was open-ended, whereas place was singular. As a design method, what was critical about site specificity was its multiple applications across all scales and design disciplines. Furthermore, site specificity itself was a concept that was open enough so that it evolved, expanding its range of application, as Craig Owens and Miwon Kwon have said, from site itself (as highly specific readings of a physical location) to the institutional frame (as cultural and political site of a work's production), to discursive site (that is, gender, nature, or sexuality as sites themselves).[4] It is in both of these senses—the tangible and the conceptual readings of site—that site-specificity has been a productive model for landscape architecture during the last twenty or more years. From gardens to parks, to building surfaces, to walks, to provisional installations, site-specific work in landscape architecture has, as in art, produced a wide range of modes of expression and engagement with place.

Process has also had, like sitespecificity, its own evolution and broad scope of meanings. Process engages the dynamic condition of landscape—living material that changes over time—as the fundamental basis of design: the materials, forms, and character of a landscape reflect the processes of its making. Ecology is fundamental here, as a set of contingent and not fully predictable relationships between organisms. The capacity of landscape to make itself—its productive agency, nature making nature—is the broadest and oldest use of the process idea. Process in this sense is technique, a way of understanding and articulating a project in terms of its material determinants.

Process is also engaged for aesthetic and phenomenological effects, to stimulate a subjective engagement with landscape, one that is renewed as changes in color, texture, spatiality, and scents unfold through the seasons and, over the long term, through the changes brought on by the landscape's growth and decay. A corollary to this conception of process is time as a narrative embedded in the medium of landscape and thematized through various measures of change on the site. More recently, process has expanded beyond the ecological and the phenomenological to the programmatic and the social. In this sense, process refers to changes in the social uses of a park due to shifts in population, emergent trends in recreation, community participation, and the incorporation of an ever-growing diversity of cultural practices into public landscapes. All of these bring changes to the site that affect not only its organization but the experience it offers to its users.[5]

These three different ways of understanding process did not arise simultaneously. Process has acquired specific meanings and connotations as design practice critically evaluates and advances the discipline. At the Bos Park in Amsterdam (built between 1929 and the 1950s), for instance, process provided a framework for breaking away from pictorial approaches to design. In the work of Hargreaves Associates of the 1980s, it provided a critique of static, internalized, modernist composition and a way to reintroduce a subjective dimension of landscape that had been repressed by the overly positivistic designs of the previous two decades. More recently, the work of OMA/Rem Koolhaas and West 8/Adriaan Geuze during the 1990s engaged process to redirect the rational demands of a project toward a creative end. Logistical issues such as zoning and program became driving forces for innovation and transformation in design.

Working with a process-based approach, rather than a purely compositional one, demands four significant shifts in design methodology. First, the dynamic nature of the material itself requires one to design processes rather than a landscape's final form. Instead of introducing external forms and transforming the site to accommodate those forms, these are "found" and evolved out of systems already there. This implies a shift from creating compositions based on notions of balance, regularity, and hierarchy to working with systems, natural or man-made, and the various ways in which they can be organized and distributed as fields, gradients, matrices, corridors, etc., to facilitate connectivity, ecological functions, program, and the perception of phenomena.

Second, there is a shift in design methodology toward dedicating more effort to site research than once was the case in formally focused design approaches. Thus in addition to the standard ecological inventory, site research includes a broader set of concerns that extends beyond property limits, such as economic interests, demographics, migration patterns, politics of resource allocation, and toxicity. Site research also explores how systems have evolved and performed over time, questioning how and why the landscape arrived at is present state, in addition to registering what is already there.

Third, history is understood as a process itself, rather than a visual reference for form, style, or type. Process-based practices acknowledge that the site is defined as much by its visible physical qualities as by its accumulated histories. This is especially relevant to large parks because they occupy sites that have been transformed several times over the course of centuries. For instance, the first European parks were typically royal hunting grounds, large tracts of forested land that often included swamps or wetlands where wildlife congregated. Others, such as those in Rome, were large villa estates that had been agricultural land, and often these were ancient forests or burial sites before they were cultivated. Early American parks of the mid-nineteenth century were usually sited on land not ideally suited for development or outside of the city. Contemporary parks in Europe tend to be sited on abandoned industrial sites, some of which were previously agricultural, while in the United States recent proposals for large parks have involved old airports, abandoned military bases, and post-closure landfills. Needless to say, the natural history of the site—those larger ecological patterns and systems such as geomorphology and hydrology—is always a determining factor. Therefore history is a way of understanding the many forces at work on a site. "Existing condition" plans are expanded to include information on a site's formal structures, but also to reveal a site's trajectory toward its present condition. What was it before it became a hunting ground, a steel mill, an agricultural field? What are its geologic origins, and how have patterns established by geology been transformed, or made to remain legible, on the site? Which are the persistent qualities of the topography, vegetation, and drainage? What has adapted to change? What hasn't? What are those external events, in economics, politics, and environmental regulation, that affect the site and have given impulse to its development?

Fourth, process-based practices anticipate change from the outset, understanding that their intervention is only one of many in the immense evolutionary process of the landscape. Design in this case is less about permanence and more about anticipating and accommodating growth, evolution, and adaptation in the face of unexpected disturbance and new programs and events. As a result, more weight is placed on establishing an argument for the objectives of a project than on creating a vision for a final form. And the critical evaluation of the project takes into account the types of research, the scenarios it considers, and the frameworks for adaptive change it sets out as much as the expressive qualities of its systems.

Design Strategies for Large Park Sites

Although the complexity of large parks demands a process-based design approach, landscape architects have developed strategies that direct process toward the expression of place. Building memorable places remains a driving ambition and fundamental value of the discipline, in spite of the emphasis on process-related issues in recent practice. These strategies, however, reveal a notion of place that is broad and complex. An inclusive attitude toward history, ecology, recreation, and perception has transformed place from inert visual scene into an historically contingent process always in a state of formation. This is expressed through design strategies that layer multiple modes of organization and establish a range of dynamic processes on the site, from open to closed, that leave the possibility for unexpected events to unfold. The creative interpretation of program engages the notion of place as social space. Finally, bridging scales, from the vast to the intimate, and engaging all the senses, address the inscription of the body in place.

Organization

Layering multiple forms of organization on the site is a strategy that acknowledges complexity, history, and the often contradictory programs that must be accommodated in large parks, without subsuming them under a single language of design. The various modes of organization are conceptualized as independent of each other, superimposed on the site so that although they may intersect, they do not add up to a unified aesthetic whole. Typically they remain distributed in a nonhierarchical way through the surface of the site. This method of generating and organizing the plan, although well known in the recent past, can also be detected in European parks of the late nineteenth century, especially in public parks developed from royal hunting grounds. As these sites became incorporated into the expansion plans of the city, they were integrated into networks of urban infrastructure. This is most evident in the extension of the circulation patterns, which shift scale for different types of connection: those that are external and serve to connect to the city and those that remain internal to the park. At the Bois de Boulogne in Paris (established as a royal park in an ancient forest in 1527, made into a public park in 1852), a northeast-southwest drive cuts through the park to connect surrounding neighborhoods to the Avenue Foch and the Champs Elysées, both key avenues in Haussmann's plan for the expansion of the city (FIG. 1). In addition, a north-south avenue through the park connects Neuilly to Boulogne, and several more localized roads connect major programs to adjacent neighborhoods, integrating the 2,090-acre site into the city. Adolphe Alphand overlaid a network of sixty-one miles of sinuous pathways that engage the internal spaces of the forest and gardens, independent of the city in form and patterns of connectivity. Similarly, at the Tiergarten in Berlin (a royal hunting ground made into a public park in 1740, redesigned by Peter Joseph Lenne between 1832 and 1839), infrastructures of circulation and connection clearly integrate the park into the larger organizational system of the city, and respond in their scale to spaces outside the park. Superimposed and within these large-scale connections are smaller localized networks that take on a different form—nongeometric, nonpragmatic, nonfunctional—that have to do with movement and experience within the spaces of the park (SEE PAGE 205, FIG. 5).

In their competition entry for Downsview Park of 2000, James Corner and Stan Allen integrate the park into the city by proposing a topographical strategy that facilitates movement through the site. Different types of flows cut through and join park, city, and region: a through-flow ecology of wildlife, a swale system that is an extension of the ravines outside the site, and a network of meadowways that connect to the surrounding regional ravine and woodland system. These contrast to a closed-circuit ecology for program and events that remains internal to the site (FIG. 2). Whether to integrate an existing landscape into the expansion of a city or to generate a plan for an isolated and unarticulated site, proposing networks as a series of independent superimposed layers is a strategy that, while giving the site logic and cohesion at a large scale, also facilitates the introduction of variety of spatial conditions and programs at a local scale. Layering diverse modes of organization has been a well-known design operation since Bernard Tschumi's and OMA/Rem Koolhaas's proposals for the Parc de la Villette in 1984, which have received extensive coverage in design literature. But as Elizabeth K. Meyer demonstrated in 1991, this had been a design strategy for urban parks since the nineteenth century.[6] The fundamental difference between those layers designed by Tschumi and the ones designed by Corner and Allen, as well as James Corner/Field Operations in their proposal for Fresh Kills, and in their work generally, is that they are specific to the particular ecologies, programs, and

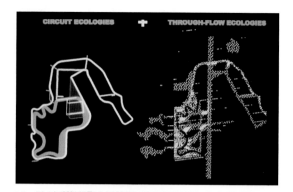

FIG. 2: James Corner + Stan Allen et al., "Emergent Ecologies," Downsview Park, Toronto, diagram of organizational systems (top).

FIG. 3: Parc du Sausset, Villepinte, illustrative plan drawn by Anna Horner, 2003 (bottom).

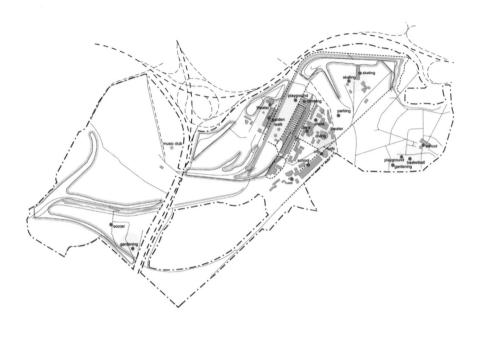

FIGS. 4, 5, 6: Landschaftspark Duisburg-Nord, topography and program diagram drawn by Gina Ford, 2003 (top); view of park, 1999 (above, left); and view of water garden, 1999 (above, right).

site conditions of the place. In addition, the layers are dynamic—that is, they are meant to change in size, shape, plant composition, and even function as they adapt to changing conditions over time.[7]

Superimposing different modes of organization is also a strategy to reveal the multiple histories of a site. The Parc du Sausset (494 acres, designed by Michel and Claire Courajoud, 1981) was constructed on agricultural land on the periphery of Paris to serve the recreational needs of a largely immigrant population. Cultivated for centuries, the site had once been forested. At the time the project began, the site was empty and largely flat, with rich soils that were conserved for the project. The Courajouds' intention was to recover the region's loss of cultural identity by reintroducing a forest and agricultural land into the barren site.[8] The design includes a 173-acre forest, 272 acres of lawns and meadows, a 5-acre marsh, and a 12-acre pond built at the same time as the surrounding housing. The distribution of these elements on the site can best be described as a gradient, from densely forested toward the northwest to alternating clearings and forest, to enclosed meadows (*bocage*), to formal lawns and gardens toward the southern part. This gradient of landscape types generates diverse spatial conditions (closed, semi-open, open) that are unified by the long and continuous gestures of circulation, by the water cycle of river/ wetland/lake that runs through the southern half of the site, and by the landforms that maintain a continuous ground over the infrastructure that cuts through the park (FIG. 3).

Peter Latz's design for Landschaftspark Duisburg-Nord accepts the steel production plant largely as it is, forming the underlying base into which he strategically inserts new programs and gardens (FIG. 4). Machines and storage spaces are colonized in creative ways to accommodate new programs and social activities, giving the impression that the park is a place of multiple identities, from the industrial sublime to the bucolic pastoral, to an intensely programmed active sports ground. This effect is in part achieved by accepting existing structures and reprogramming them to new uses, and partly by an administrative structure that has multiple owners, tenants, and agencies responsible for different aspects of the park's funding and maintenance. This strategy is powerfully reflected in the ground of the park—a surface with no paths, a place of dispersal with no directionality, hierarchy, or structuring narratives. The park feels less like something with a cohesive identity and more like a free zone, an ambiguous territory for roaming around in search of events, encounters, and visual experiences. Thus while retaining the history of the site as image and palimpsest, and embracing the diversity afforded by the new administrative structure, Latz + Partner leave the connections between past and present unresolved (FIGS. 5, 6).

Both the Parc du Sausset and Landschaftspark Duisburg-Nord combine a retrospective with a projective attitude toward place. The designers retain and recover traces of different historical periods without privileging the present at the exclusion of the past, and vice versa. They maintain aspects of the site's

FIG. 7: Bos Park, Amsterdam, illustrative plan drawn by Rebekah Sturges, 2003.

history, but they do so without intentions of reconstructing the past. To the contrary, they recast old forms into new uses, projecting a future onto the site while addressing the continuity of its physical structures. Large parks inevitably build on existing places. To the extent that landscape architects creatively engage the traces left by previous uses, the parks become unique and memorable places.

Open (Dynamic) and Closed (Formal) Systems

As mentioned previously, process in landscape architecture typically refers to the use of nature's productive capacities as technique to develop the landscape. This is often a necessity in large parks, whether to accommodate budgetary constraints, remediate soils, or achieve desired aesthetic effects. Not every element of a park is left open to process, however, and it is often in deciding what is to remain open-ended in a large site that aspects of place are revealed.

At the Bos Park, for example, water is essentially a closed system (FIG. 7). The series of canals that traverse the site work as a long, narrow lake. Water is released into the Nieuwe Meer using a pump, but only after rain has raised the level of the water in the canals sufficiently to cause flooding in the adjacent forest and meadows, which rarely occurs. The planting of the forest, on the other hand, engaged successional processes for its development and remains open-ended. The designers used a well-established practice of industrial forestry that depends on plant succession. Randomly distributed in a grid, two forest types were planted: a provisional pioneer forest of fast-growing alders, willows, poplars, and birch, and a permanent forest of the slower-growing ash, maple, oak, and beech. In addition to helping drain the site through evapo-transpiration, the pioneer forest provided shelter for the seedlings of the permanent forest, which require shade during their initial years. After fifteen years of growth, the pioneer forest, with the exception of the alder, was cut down, allowing the established permanent forest to grow. The remaining alder was then pollarded to produce a horizontal branching habit and provide shade on the forest floor to prevent understory growth.[9] Within this underlying matrix of trees, however, a series of stable objectlike spaces and landforms create a counterpoint to the experience of the forest. The hill (an atypical formation in this landscape, as is the forest) was made from the soil dug for the canals and the lake, and has a distinct profile. Equally conspicuous is the 1.2-mile-long rowing course, rectangular in shape and sited perpendicular to the direction of drainage flow, immediately distinguishing itself visually and functionally from the other water bodies on the site. The open lawns are mowed and not allowed to go into succession, remaining closed to external stimuli, stable in their visual and programmatic intentions (FIGS. 8–10).

A similar mix of open and closed systems is in effect at the Parc du Sausset. As in the Bos Park, the forest was started as seedlings and allowed to grow, but as a single, one-time planting, and without much succession

FIG. 8: Bos Park, Amsterdam, view of forest open to succession, 1999 (top).
FIG. 9: Bos Park, Amsterdam, view of formed space, 1999 (bottom).

FIG. 10: Bos Park, Amsterdam, view of meadow and lake, 1999.

allowed. The lawns and meadows in the park receive different management, from frequently mowed (in the formalized lawns for picnics and games) to open to succession (as in the meadows enclosed by hedgerows in the bocage). The water sequence is also open, as it has a role in stormwater cleansing, in addition to providing habitat.

The Mark Rios/Roger Sherman proposal for Fresh Kills, called "rePark," engages that site's previous function as landfill in both direct and analogous ways. The overarching concept for the scheme is that of evolution and change. A dual strategy augments and diversifies existing conditions. First, the eight ecologies that organize the ground are habitats found on site and are projected to evolve without intervention.[10] Then, superimposed on this mosaic are the transects, conceptualized as bands of provisional events and programs that traverse the mounds in trajectories of varying locations and length. Analogous to the landfill itself, the transects are accumulations of things made elsewhere. They are collectively designed, unlike traditional plans, and include contributions by landscape architects and artists not on the initial team. As such, the proposal foresees but does not script a series of permutations over a long period. In terms of park management, the design proposes a maintenance regime like the one at work for the landfill. The park is to be maintained as necessities arise, mostly as localized applications within the transects and other areas of intense use (FIGS. 11, 12).

The idea of process that comes out of James Corner/Field Operations' Fresh Kills proposal is that landscape is a performative, reactive, and dynamic medium. The scheme orchestrates all of the site's histories—freshwater marsh, salt marsh, landfill—and its future as park, by initiating a series of imaginative and fairly technical first-stage mechanisms and devices that will initiate the transformation of the site. Threads, mats, and islands, each with its own organizational strategy (linear, matrixlike, free-standing masses), respond to the specific site conditions. Linear forests, forming threadlike systems of

EXISTING MONITORING FIELD

EXTENDED MONITORING FIELD
(VIRTUAL MAINTENANCE COORDINATES)

POTENTIAL TRANSECT

POTENTIAL TRANSECT NETWORK

FIGS. 11, 12: Mark Rios/Roger Sherman, "rePark," Fresh Kills competition proposal, transect network evolution (top) and diagram of eight ecologies (below).

connectivity, are located on the northeast slopes of the landfill, to take advantage of ground moisture and shelter from prevailing winds. Mats comprise the predominant surface areas and are distributed throughout. Islands are envisioned as clusters of vegetation located in the south and west slopes, where dry soil and exposure to winds make difficult the establishment and long-term viability of extensive plantings. In addition to being a response to site conditions, however, each of these types is given a performative task: threads are distributive, containing, in addition to linear forests, drainage swales as well as links and circulation; mats are colonizing, extending throughout as sports fields and event areas, but also as the interstitial and underlying medium that salt marshes and freshwater wetlands constitute on this landscape; islands are clusters of intensity, both ecological (for protected habitat) and programmatic (architectural structures). As distributors, colonizers, and condensers, the threads, mats, and islands are generators of further complexity.

These four projects present a wide application of the use of process in the construction of large parks. None of them, however, result in bland naturalistic landscapes. Process results in place when it is paired with additional conceptual frameworks, whether cultural, site-specific, or phenomenological, that transform it from mere technique to legible design language.

Program and Events

"To compose today means to create programs. We invent or propose them; we mix them, give them support, denaturalize them.... Programs are mutable, transformable in time. We must define programs which can forget or can be transformed later."[11] With this definition Federico Soriano captures the primacy of program in architecture and landscape architecture over the last three decades.[12] If the material histories of the site inform the ways in which the previous two strategies (layering diverse forms of organization and introducing a range of dynamic systems) embed specificity of place into large parks, program is another fundamental strategy, as generative of organization and form as is site history, to create place. Large parks become meaningful places as much by their physical qualities as by the events that take place within them. I want to focus less, however, on unprogrammed space that is flexible and open to multiple events (the widely accepted "programmatic indeterminacy") and more on how event can engage the particularities of a site in material and cultural ways, augmenting the specific qualities of a place.

In the Mathur/da Cunha + Tom Leader Studio proposal for Fresh Kills, titled "Dynamic Coalition," the notion of program is tactical, the very thing that generates the park. Based on the idea that the park is a site of many depositions—crushed rocks, glacial till, marsh detritus, garbage, and World Trade Center debris—the events planned for this proposal are understood as depositions themselves that, over time, will endow the site with particular meanings that reflect the site's geologic and cultural origins. These operate at many scales to construct the surfaces, edges, and program of the park. First are events

FIGS. 13, 14: Mathur, DeCunha + Tom Leader Studio et al., "Dynamic Coalition," Fresh Kills competition proposal, events calendar (below) and March events (opposite).

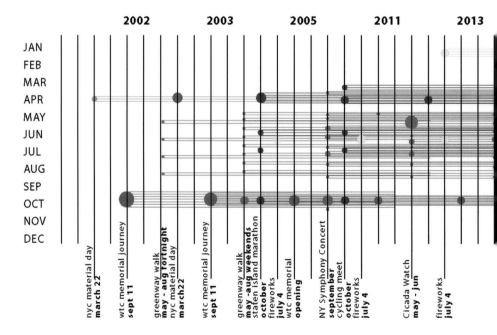

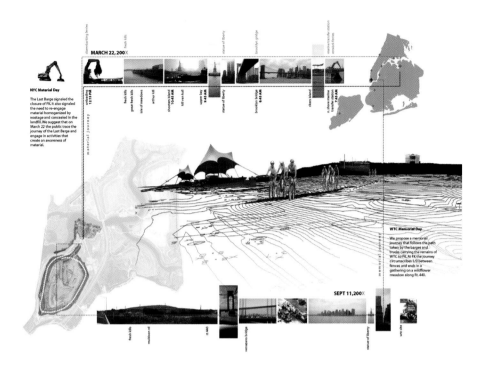

MARCH 22, 200X

NYC Material Day

The Last Barge signaled the closure of FK. It also signaled the need to re-engage material homogenized by wastage and concealed in the landfill. We suggest that on March 22 the public trace the journey of the Last Barge and engage in activities that create an awareness of material.

material journey

unloading 12:15 PM
fresh kills
great fresh kills
isle of meadows
arthur kill
shooters island 10:45 AM
kill von kull
upper bay 9:45 AM
statue of liberty
brooklyn bridge 8:45 AM
n. shore marine transfer station 7:45 AM
rikers island

WTC Memorial Day

We propose a memorial journey that follows the path taken by the barges and trucks carrying the remains of WTC to FK. At FK the journey circumscribes 1/9 between fences and ends in a gathering on a wildflower meadow along Rt. 440.

memorial journey

SEPT 11,200X

fresh kills
muldoon rd
rt 440
verrazano bridge
statue of liberty
wtc site

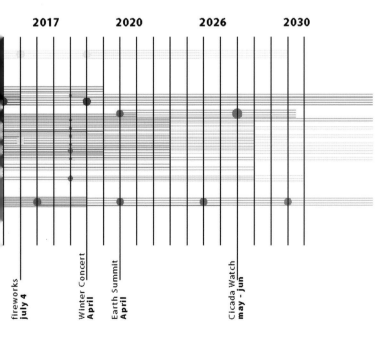

2017   2020   2026   2030

fireworks
july 4

Winter Concert
April

Earth Summit
April

Cicada Watch
may - jun

specific to this site. These events are ritualistic and take the form of two journeys intended to mark the site with meaning. One is a material journey, referring to the site's history as landfill; the other addresses the site's new identity
as memorial ground. Second are events specific to Fresh Kills although not
exclusive to it, that relate to the ecology of the site—a cicada watch, greenway walks every two weeks between May and August—and art, technology,
and science events that explore the future and the history of the waste stream.
Third are cultural events meant to draw visitors from the region, such as concerts, fireworks, and marathons. Finally, there are disturbance events. By this
I mean events that disturb present physical boundary conditions on the site
at neighborhood and regional scales and open up the park for unforeseeable
social and programmatic uses. At the neighborhood scale are the cuts on the
20-foot-high berm that separates the site from Staten Island Mall. The gaps
are developed as market areas for recycled or processed materials, thus merging commercialism with recycling. Housing is introduced on the edge along
Victory Boulevard on a reclaimed spit of land, and a neighborhood recreation
center is located on the Arden Heights Woods edge. At a regional scale, Route
440 is transformed from a single-use expressway that divides the site into a
diversified thoroughfare with the addition of light rail, bicycle path, and promenade. Together with the intermodal platforms, parking groves, and wildflower meadows, Route 440 is reprogrammed as a zone of rich and dynamic
interface that brings together separate areas of the site. A series of intermodal
stations will connect the site to I-278, New Jersey, and Brooklyn, and to local
ferries, expanding the reach of the site across the New York metropolitan area.
It is important to emphasize the nongeneric nature of these events and how,
even though they are specific to the landfill, they touch it with different inten-
sities. Some events are ephemeral and meant to disappear, others make light
imprints, and others are permanent interventions on the site (FIGS. 13, 14).

The material conditions of the landscape itself can become events
unique to the site and thus lead to the construction of place. Landscape itself,
as living medium, is event. The red and scarlet oak forest on mound 1/9 in the
Hargreaves Associates proposal for Fresh Kills, "Parklands," ties material conditions to the commemorative program of the preserve. The horizon, a unique
condition afforded by the landfill topography, is emphatically presented in this
proposal in multiple ways, as flat, inclined, rounded, empty, vegetated, solid,
and liquid, constituting a significant experience in the park. In the James
Corner/Field Operations scheme, landscape as living material is the central
event of the park. Corner pushes the limits of material conditions on the site,
creating situations in which expectations about a natural setting are destabi-
lized. For example, on the old siltation basins of the landfill, a new ground that
has changed the habitat from saltwater to freshwater wetlands, Corner proposes a sweetbay magnolia (*Magnolia virginiana*) swamp, marginally hardy in
this latitude although hardy in Staten Island because of the estuary effect on
its climate. The location of the non-native freshwater swamp next the native

FIG. 15: James Corner/Field Operations et al., "Lifescape," Fresh Kills competition proposal, freshwater wetland and saltwater wetland.

FIG. 16: Interiority in James Corner + Stan Allen et al., "Emergent Ecologies," Downsview Park, Toronto (top).
FIG. 17: Interiority in James Corner/Field Operations et al, "Lifescape," Fresh Kills competition proposal (bottom).

saltwater wetland is intended to produce a surreal quality, indexing the illogical juxtaposition of the two ecologies. This is further emphasized by the location of a pedestrian walk on the boundary between the two, by the use of the purple flowering water lily rather than the more common, native white one, and by the large gabion wall that retains the soil in the wetland (FIG. 15).

Finally, large parks often bear traces of major historical events. The Casa de Campo (a royal hunting ground turned into a public park in 1931), located on a bluff overlooking Madrid, was a strategic site for Franco's nationalist troops, who bombarded the city from the heights of the park in 1936. It was also the site of prolonged battles. Similarly, Berlin's Tiergarten was destroyed by Allied bombing during World War II. Later, the park was cut off from the rest of the city by the Berlin Wall, which allowed new species of plants to grow in the park, substantially increasing its biodiversity.[13] The Fresh Kills site contains the screened material from the World Trade Center. Reforestation and replanting, topographical changes, monuments, and contemplative spaces inscribe large parks with unique places of remembrance that mark the events of their varied histories.

Scale

The strategies explored thus far contribute to the building of difference within what can often be the unarticulated and homogeneous space of a large park site, especially if the site is being reclaimed from a landfill, an airport, or an old quarry. The recovery of material histories, the introduction of new spaces, experiences, and aesthetic qualities, and the insertion of program scaled for different kinds of recreational and social activities all contribute to the recognition that within one place there are many scales simultaneously at work. Furthermore, current conditions of global culture in metropolitan areas index place as no longer simply local but formed by a tension between the immediate physical setting and an extended network of associations. This scalar tension drives the design of the journeys of the Mathur/da Cunha + Tom Leader Studio proposal for Fresh Kills, which point to the larger context of New York City. In this sense, largeness is not just the size of the park but the dialog it has with its wider context.

But if large parks are to remain vivid experiences in the imagination of the public, they must include shifts in scale—from the vast to the physically enclosed—to activate relations between the body and landscape. Gaston Bachelard's observation that the notion of immensity is fundamentally intimate informs Corner's proposals for Fresh Kills and Downsview.[14] In these projects there is a deliberate juxtaposition of vastly different scales, whereby place can be shown to distend beyond any graspable measure, and yet retain a focus on the much smaller space of the body. For instance, at Downsview Park Corner and Allen contrast the interiority of the habitat nest with the open meadows by siting paths in the transition zone between the two, where both scales are simultaneously engaged. At Fresh Kills, the intimacy of the path as a location scaled to the body is juxtaposed to the immensity of the shifting horizons visible from the path. In these projects circulation is not only connectivity, it is strategic positioning to intensify place (FIGS. 16, 17). In addition, there is the calculated positioning of the body in locations of powerful sensory experience. Corner and Allen's habitat nests at Downsview, a series of basins that collect water and then release it into the ground, are places that engage sound, texture, and smell, in addition to sight. Using a variety of soils, as well as a grading pattern of ridge-and-furrow of different heights and depths, they introduce several ecological communities with specific habitat plantings, each with a distinctive texture, scale, and color. The boardwalks that thread through these nest habitats are designed so that they hover as closely as possible over the ground. The absence of railings on the platforms facilitates a closer body/landscape interface with water, dampness, smells, colors, and the varying spatialities of the basins (SEE PAGE 37, FIG. 3). Place is constructed in these instances as a mediating ground between the unbound landscape and the body.

Conclusion

The nature of contemporary practice generally, and of large parks in particular, has required a focus on issues unique to large public landscapes: how to engage and coordinate a tangle of multiple interests, constituencies, regulations, and economic challenges, how to incorporate change and adaptive measures over time, how to reconcile the seemingly contradictory objectives of restoration of ecologies and intensely programmed sites, etc. We do not discuss visual, spatial, or phenomenological qualities; we discuss frameworks, emergence, the performative, the adaptive. The shift from composition to process has facilitated the incorporation of complexity in design, but it has also given less visibility to issues that, in spite of new methods, remain at the core of the discipline, such as maintaining and expressing the quality of place and its cultural meaning. Embedded in the methodologies of process-based design are the very things that make places recognizable as memorable and unique: the legibility of the various forces at work on a site, the inclusion of traces left on the land by previous uses, the expression of environmental change, the accommodation of multiple scales, the commitment to diversity, and the determination to adapt existing forms to new social practices.

NOTES

1. For a discussion of process in smaller sites, see Elizabeth K. Meyer, "Post-Earth Day Conundrum: Translating Environmental Values," in Michel Conan, ed., *Environmentalism in Landscape Architecture* vol. 22 (Washington, D.C.: Dumbarton Oaks Research Library and Collection), 187–244.

2. References on place are numerous. Particularly helpful are Edward S. Casey, *Getting Back into Place: Towards a Renewed Understanding of the Place-World* (Indianapolis: Indiana University Press, 1993) and *The Fate of Place* (Berkeley: University of California Press, 1999). Yi-Fu Tuan, *Space and Place: The Perspective of Experience* (Minneapolis: University of Minnesota Press, 1976); James Duncan and David Ley, *Place, Culture, Representation* (London: Routledge, 1993); Allan Pred, "Place as Historically Contingent Process: Structuration and the Time-Geography of Becoming Places," in *Annals of the Association of American Geographers* 74, no. 2 (June 1984): 279–97; and J. B. Jackson, *A Sense of Place, a Sense of Time* (New Haven: Yale University Press, 1994).

3. Duncan and Ley, *Place, Culture, Representation*, 303.

4. Likewise, references on site-specific art are numerous. See, for instance, Robert Irwin, *Being and Circumstance: Notes Towards a Conditional Art* (San Francisco: Lapis Press, 1985); Rosalind Krauss, "Sculpture in the Expanded Field," in *The Originality of the Avant-Garde and Other Modernist Myths* (Cambridge, MA: MIT Press, 1986), 276–90; Miwon Kwon, "One Place After Another: Notes on Site Specificity," *October* 80 (Spring 1997): 85–110; Craig Owens, "Earthwords," "The Allegorical Impulse: Toward a Theory of Postmodernism," "The Allegorical Impulse: Toward a Theory of Postmodernism, Part 2," and "From Work to Frame, or, Is There Life After 'The Death of the Author'?" in *Beyond Recognition: Representation, Power, and Culture* (Berkeley: University of California Press, 1992), 40–51, 52–69, 70–87, 122–41. For site-specific work in contemporary landscape architecture, see Christophe Girot, "Four Trace Concepts in Landscape Architecture," in James Corner, ed., *Recovering Landscape: Essays in Contemporary Landscape Architecture* (New York: Princeton Architectural Press, 1999), 59–67; Sébastien Marot, "The Reclaiming of Sites," in *Recovering Landscape*, 45–57; Elizabeth K. Meyer, "Site Citations: The Grounds of Modern Landscape Architecture," in *Site Matters: Design Concepts, Histories, and Strategies* (New York: Routledge, 2005), 93–129.

5. Examples of twentieth-century projects that engage process as technique of construction are: the Bos Park in Amsterdam by Van Eesteren et al., and the work of Michel Desvigne and West 8/Adriaan Geuze. For process and changing experience on the site, see much of the recent work of Michael Van Valkenburgh and Associates, such as the ice walls, Mill Race Park, and Teardrop Park (2005); likewise, much of the work of Hargreaves Associates, such as Byxbee Park and Louisville Park; and Michel Desvigne's residential work. For process and program, see the OMA/Rem Koolhaas project for La Villette (1984) and, with Bruce Mau, their project for Downsview Park "Tree City" (2001); and West 8/Adriaan Geuze, Theatre Square Rotterdam.

6. Elizabeth K. Meyer, "The Public Park as Avant-Garde (Landscape) Architecture: A Comparative Interpretation of Two Parisian Parks, Parc de la Villette (1983–1990) and Parc des Buttes-Chaumont (1864–1867)," in *Landscape Journal* 10, no. 1 (1991): 16–26.

7. Consider the differences between the through-flow diagrams for Downsview, and the threads, mats, and cluster diagrams of the Fresh Kills proposal.

8. Quoted by Anna Kaufman Horner in the exhibition Large Parks: New Perspectives held at the Harvard University Graduate School of Design, April 2003.

9. Anita Berrizbeitia, "The Amsterdam Bos: The Modern Public Park and the Construction of Collective Experience," in Corner, ed., *Recovering Landscape*, 187–203.

10. Rios Associates, "rePark," *Praxis: Journal of Writing and Building* 4 (2002): 49.

11. Manuel Gausa, Vicente Guallart, Federico Soriano, et al., *The Metapolis Dictionary of Advanced Architecture* (Barcelona: Actar, 2003), 499. The same has been stated by Bernard Tschumi in *Event-Cities* (Cambridge, MA: MIT Press, 1994), 11: "There is no architecture without action or without program"; and Alex Wall, in "Programming the Urban Surface," in Corner, ed., *Recovering Landscape*, 243: "The space of form is here replaced by the space of events in time."

12. See, for instance, Jacques Lucan, "Parc de la Villette," in *Rem Koolhaas/OMA* (New York: Princeton Architectural Press, 1996), 86–95; Bernard Tschumi, *Event-Cities*—especially his definitions of crossprogramming, transprogramming, disprogramming, and programmatic dissociations; Wall, "Programming the Urban Surface," 233–49.

13. Caroline Chen, in the exhibition Large Parks: New Perspectives.

14. See Gaston Bachelard, *The Poetics of Space*, trans. Maria Jolas (Boston: Beacon Press, 1969), 194.

FIG. 1: Pedestrians and joggers in Chapultepec Park, Mexico City, recently renovated through a public-private partnership.

CONFLICT AND EROSION:

THE CONTEMPORARY PUBLIC LIFE OF LARGE PARKS

JOHN BEARDSLEY

It is increasingly difficult to find a large park anywhere in the world that is fully public—that is, entirely free and accessible in all places at all times and fully supported by public funds. To some extent, this is a function of their physical scale and social complexity. Large parks are more difficult than small ones to finance and maintain, which has resulted in the growth of public-private partnerships to manage their construction and upkeep. They are characterized by greater ecological complexity than small parks, along with more diverse and dispersed programs, all of which put significant management pressures on their public sponsors. At the same time, large parks serve more diverse constituencies than small ones. While the latter might be situated within coherent and even homogeneous neighborhoods, the former are typically adjacent to several different communities and necessarily serve more diverse socioeconomic groups. These circumstances are beginning to have an impact not only on the way parks are managed but also on how they are designed: the existence of multiple constituencies has led both to a focus on adaptability—to accommodate different user groups at different times—and to designs that are visibly if somewhat awkwardly divided into distinct precincts to serve various publics. Hargreaves Associates' recently completed Chrissy Field in San Francisco, for example, though not an especially large park, nevertheless gives pointed evidence of such parcelization, with distinct areas for windsurfing, bird watching, and dog walking. The windsurfers got beachfront parking, the dog walkers got a large field for active recreation, and the birders got a wetland fenced to protect them and their quarry from the dogs. Suffice it to say that we can no longer assume a singular public; we have to talk of multiple, often conflicting, publics in the plural. Similarly, we cannot talk simply of public as opposed to private space—we have to acknowledge a blurring of the distinctions between them.

Indeed, growing privatization may be the greatest challenge facing large urban parks, signaling an erosion of commitment among public institutional sponsors to capitalize and maintain them. In the United States, private groups are increasingly responsible for the management of parks. One of the first and still most familiar of these groups is the Central Park Conservancy in New York, established in 1980 to remedy the deteriorating physical condition of the park and its dwindling budgets for maintenance and recreation. The Conservancy experiment was wildly successful: by the late 1990s, the organization had a staff of 250, some 1,200 volunteers, and an endowment of $65 million.

It was so successful, in fact, that in 1998 the city of New York contracted with the Central Park Conservancy to take over most of the day-to-day maintenance of the park. Similar public-private partnerships in park management have emerged in other cities. In Boston, private support for public parks began with the organization of Friends of the Public Garden in 1971 and the Franklin Park Coalition in 1975. Several years later, these groups joined with others to form the Boston GreenSpace Alliance. In Baltimore, three private groups— the Urban Resources Initiative at Yale, the Trust for Public Land, and the local Parks for People Foundation—have been instrumental in acquiring land and commissioning a design for the Gwynns Fall Greenway, which connects suburban Leakin Park with Baltimore's inner harbor, running through some of the city's poorest neighborhoods. These public-private partnerships, though highly commendable in themselves, have some troubling public policy implications—to which I want to return.

Such partnerships are only one of the ways in which the line between public and private is being blurred. Growing numbers of private concessions are being granted within existing parks: for instance, at the Bois de Boulogne, a 2,100-acre park in Paris, many of the public open spaces originally part of the park, such as a skating pond, have become private sporting clubs that are off-limits to all but their members. There are private tennis and polo clubs in the park, a paid amusement area, and a 17-acre campground with seventy-five mobile homes for rent.[1] New parks are following the same pattern. When an abandoned Thyssen steel mill in Germany became the 620-acre Landschaftspark Duisburg-Nord, some of the unofficial occupations of the derelict space were formalized, gentrified, and privatized. People had been using unrestored cooling tanks for scuba diving and empty buildings for raves; now, you pay to join an approved diving club or to enter a sanctioned nightclub.[2] On the positive side, unofficial community gardens already in existence on the site became part of the accepted uses of the park; and new bike paths, athletic facilities, and workshops for vocational training were added to ensure that the park would be relevant and accessible to a wide variety of publics.

That this blurring of public and private is something of a global phenomenon is suggested by recent evidence from Mexico and China. In Mexico City, the landscape architect Mario Schjetnan and his firm, Grupo de Diseño Urbano, have been involved in the restoration of Chapultepec Park, the most extensive and historically significant landscape space in the central city (FIG. 1). The park is visited by 15 million people per year, many of them drawn to the zoo and the Anthropology Museum, which lie within its precincts. Heavy use and years of neglect had degraded the park: the forest was overplanted and overgrown, planting beds were abandoned, lake water was badly polluted, and the park was overrun by cars and unauthorized vendors. The first phase of restoration, completed in 2005, consisted of improvements to 300 of the approximately 600 acres in the park's central section. Trees were thinned and underbrush removed, allowing grass to reappear beneath the

FIG. 2: Temple of Heaven, Bejing. Local residents use the outer areas of the park for exercise, games, and music, while tourists visit the inner precincts.

canopy; the primary entrance to the park was restored; display beds and gardens were replanted; vendors were relocated to food courts and concession areas; and numerous infrastructural upgrades were made, among them improvements in water treatment, lighting, and irrigation. Much of the work was accomplished using private donations. According to Schjetnan, "The mayor said that for every peso you raise privately, I will give you a peso."[3] Citizen and donor groups were established to raise the money, much of which was collected in small denominations from subway stations. This correlates with the park's demographics: 65 percent of its users come to the park by subway.[4] But it also shifts some of the burden of financing the park to those who are least able to pay.

Construction funding produced a public-private hybrid in Mexico City; in Beijing, it has been created by admission charges. The 660-acre grounds of the Temple of Heaven, for instance, a Ming Dynasty complex, are highly valued as both a local green space and an international tourist destination. Although everyone must pay to enter, there are different levels of public access marked by several thresholds. The outer area of the park, open to all at a nominal fee of about 25 cents, is the place where people of the neighborhood come to play games, exercise, and perform selections from their favorite operas (FIG. 2). The inner area of the park, containing some of Beijing's most notable historic architecture, is primarily for tourists, who pay a higher fee. The fee structure in this park underscores a more general idea that there are different levels of publicness, with access to facilities determined by the capacity to pay. From New York to Beijing, we can no longer speak simply of public as opposed to private space; we are increasingly dealing with a hybrid that is not entirely one or the other.

The erosion of publicness is linked to more subtle and even pessimistic shifts in the analysis of public space. In the view of writers such as Don Mitchell and Richard Van Deusen, for example, who contributed a chapter to the recent CASE publication on Toronto's Downsview Park competition, all open spaces are exclusionary. "The history of public space," they write, "is the history of struggle over who shall have access to it." The ideology and practice of public space in Western cities, they argue, has brought us to the point where public space is "largely depoliticized and highly controlled." Basing their conclusion on similar arguments by Michael Sorkin, Mike Davis, Nancy Fraser, and Raymond Williams, among others, they insist that "most of the open space that is planned in modern Western cities—parks, plazas, shopping malls, arcades—is decidedly not public: its purpose is to control and direct social interaction, to police it, rather than to provide a stage on which various publics can come together in all their often contentious differences and spark a conflagration of public, political, and social interaction."[5]

Most commercial spaces are decidedly not public—shopping malls, with their bans on political demonstrations, are the most flagrant example. And, as noted, parks themselves are increasingly privatized and increasingly managed by public-private partnerships. But there is a great deal of evidence that people still make relatively free use of parkland. Although there are indeed struggles over access to public parks, the situation may not be as dire as Mitchell and Van Deusen suggest: parks elude at least some of the controls more evident in commercial space. As Michel de Certeau argues, "Stories about places are makeshift things.... Heterogeneous and even contrary elements fill the homogeneous form of the story. Things *extra* and *other*...insert themselves into the accepted framework, the imposed order."[6] Things "extra and other" are always appearing in parks. Indeed, parks are one of the most reliable sites we have of unscripted social interactions. They are precisely what Mitchell and Van Deusen say they are not: stages on which various publics come together in all their contentious differences, sparking a conflagration of public, political, and social interaction. It is precisely on the subject of the contentious, even oppositional occupations of parks that recent studies conducted by graduate students in landscape architecture at the Harvard University Graduate School of Design are most interesting. I will cite a few examples.

Especially in Europe, a recognition—if not an acceptance—has developed of what might once have been considered illicit occupations. Such new uses prompt correspondingly new regulations and forms of surveillance—and new forms of evasion of the rules. In Berlin's Tiergarten, for instance, there is an area reserved for nudists. They even have a shower, donated to the park by a Herr Jahnke, a devotee of what is called *Freie Koerperkultur*. Herr Jahnke left 250,000DM to the Tiergarten on the condition that they provide a

FIG. 3: The Tiergarten is the site of Berlin's Love Parade, which overwhelms park facilities.

public shower in the vicinity of the Hofjagerallee. The Tiergarten is also the site every summer of the notorious Love Parade, which officials say turns the park into a large public restroom for a day (FIG. 3). The Tiergarten has a long tradition of public disregard for rules: soccer and barbequing are two of the most popular pastimes in the park, although they are supposed to be restricted activities. Flaunting of park rules began in the 1970s when people began occupying restricted lawn areas for picnicking and sports. In response to this pressure, park maintenance subsequently opened grass areas for public use, even allowing barbequing in one small area near the Zeltenplatz. This area constitutes only 11 percent of the park, but barbequing continues to be more widely dispersed: on a summer weekend, visitors can be seen barbequing in all areas of the park, with the Tiergarten sending out large plumes of smoke (FIGS. 4, 5).[7]

The Bos Park in Amsterdam also has a sunbathing meadow (*Zonneweide*) reserved for nudists. Here too there is evidence that rules are being flaunted. Masturbation is not permitted, but is evidently a frequent activity nonetheless in the meadow, which is colonized by a male population. The rangers say there is little crime or controversy surrounding this space, but report that there is a nudist who feels it is his duty to inform on the masturbators; evidently, even the rangers find his constant watch an irritation.[8]

The homeless and sex workers represent other aspects of unofficial occupation. By the middle of the 1990s, as many as 1,000 people had set up semi-permanent encampments in San Francisco's parks, chiefly Golden Gate. According to Peter Harnik's book *Inside City Parks*, a low-level conflict

FIG. 4: Barbequing occurs throughout the Tiergarten, although officially it is restricted to specific areas. (above)
FIG. 5: Tiergarten, Berlin, illustrative plan by Caroline Chen, 2003 (opposite).

simmered between the homeless and the city for several years, with the city complaining of trash, trampled shrubs, and illegal activities in the park, including drug use and fencing of stolen goods. When the parks department tried to enforce a curfew by reprogramming sprinklers to turn on in the middle of the night, sprinkler heads were vandalized. Conflict literally erupted in flames in November 1997, when arsonists set a huge fire in the park that required seventy firefighters and twelve trucks to extinguish. Mayor Willie Brown retaliated by ordering the camps shut down.[9]

A more complex standoff between the homeless and park authorities has been developing in Rio de Janeiro. There, Tijuca Park—the world's largest urban forest and largest replanted tropical forest—has seen the proliferation of unsanctioned settlements or *favelas* on its steep slopes and ridge tops (FIG. 6). The process has a long history: it began in the early twentieth century as a result of slum clearance in the city center and has continued unabated as increasing rents and the influx of the rural poor have driven people to Rio's "free frontier" in the Tijuca Massif. By 1993, nearly a fifth of Rio's population was living in some 545 favelas; today, there are forty-six in the Tijuca Massif alone, home to about one-third of the city's *favelado* population (FIGS. 7, 8). These settlements threaten the city as much as the park; loss of vegetation and subsequent erosion sends silt and boulders cascading down into the city every rainy season. A particularly destructive episode occurred in 1988, when massive mud slides destroyed drainage systems, blocked canals and roads, and left three hundred dead, a thousand injured, and thousands more homeless. These informal settlements are only one of the massive pressures facing Tijuca Park, which receives 1.5 million visitors a year.[10]

Why does the city tolerate these settlements? One explanation is bureaucratic: many overlapping federal, state, and local agencies administer the park; this makes it nearly impossible to determine who is responsible for park policies. But the second and larger issue is political. With a fifth of the city's population living in favelas, politicians gain a significant constituency by promising to leave the settlements alone. Moreover, there is little money or space for providing legitimate housing; illegal settlements are perceived as an expedient solution to urban growth. Some favelas have even received electricity, plumbing, and other infrastructure, transforming them into more substantial settlements. Indeed, Rio has one of the world's most interesting slum upgrade projects: called the *favela-bairro* program, it aims to transform shantytowns into neighborhoods with a minimum of physical or social disturbance.[11]

As for prostitution, the wilder and more densely forested areas of the Bois de Boulogne are heavily used for the purpose year-round—although sex workers were temporarily driven into the open in the wake of a hugely destructive storm in December 1999, which felled 25,000 trees. A park official reported that the city tolerates this activity in the Bois, as they prefer prostitution hidden in the park rather than promenaded on the boulevards (FIG. 9).

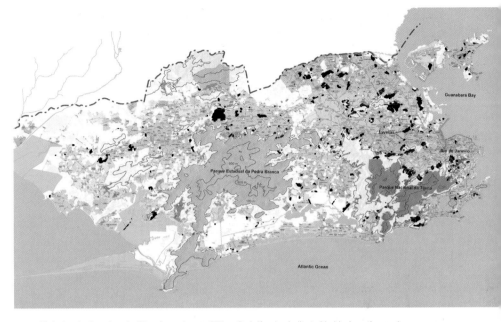

FIG. 6: Rio's *favelas* have invaded the steep slopes of Tijuca Park (favelas indicated in black on the map).

FIGS. 7, 8: Rio's *favelas*, general view of a settlement (above, left) and close-up of an intersection (above, right).

FIG. 9: Prostitution is common in the forested areas of the Bois de Boulogne.

The story is similar in Madrid, where prostitutes have been forced out of the city into a network of roads and parking lots in the Casa de Campo.[12]

These examples confirm the observation of Mitchell and Van Deusen that the struggle for control of and access to public space is far from over. There will doubtless always be conflicts between sanctioned and unsanctioned uses, between planned programs and the oppositional purposes to which parks are sometimes put. Though the abandoned steelworks at Duisburg have been upgraded into the Landschaftspark Duisburg-Nord, the park still has its more remote corners, the site of drug use and unofficial raves that attract hundreds of local kids on most summer nights. The park continues to be the site of social unrest, of which we might take the occasional presence of racist and xenophobic graffiti as an emblem. Indeed, it is possible that the conflict over this particular landscape is intensified by the somewhat sanitized history presented in the central areas of the park. Despite the acceptance of the site's gritty manufacturing history as embodied in its preserved industrial artifacts, there is no mention in park signage or literature of the close connection between the steel industry and the Nazi party during World War II, nor of the widespread use of prisoner of war and slave labor in German factories in those years. Nor is there any acknowledgment of the hostility faced by Turkish labor imported to work the mills in the 1970s. Perhaps it is beyond the range of design to address such complex social history, but we should not be surprised when conflicts erupt in places with turbulent pasts. Landschaftspark Duisburg-Nord remains a contested landscape, no matter how gentrified and sanitized it might appear in its central areas.

The evidence of social disturbance encountered in various park landscapes suggests that safety considerations will continue to be summoned as a rationale for exclusion and restraint. Moreover, there are bound to be continued disparities between the typically elite language of design and the popular culture of surrounding communities. But, contrary to Mitchell and Van Deusen and based on the evidence accumulated by Harvard GSD students in their studies of large urban parks, I would argue that such spaces are still one of the principal places where people are least subject to social control—that parks are among the few places where people are most free to pursue the ordinary and extraordinary expressions of everyday life. As distasteful as some of the conflict is that surfaces in parks, its presence seems to be one of the conditions of public life in increasingly divided societies.

As elements of larger constellations of urban public space, parks are also implicated in broader policy questions. For example, the political will and economic means needed to build large parks seem increasingly available only within the context of large infrastructure or environmental restoration projects. As more private recreational activities become available to the middle class, including the entertainment landscapes of theme parks and shopping malls, the appetite for financing large public open spaces may further decline (although commercial space and parks are increasingly conflated). Moreover, such large parks as do get built may be representative of uneven development: concentrations of capital expenditure that benefit one area or constituency at the expense of others. This was a charge leveled by Rosalyn Deutsche against Battery Park City in New York, which she thought monopolized development capital in lower Manhattan at the expense of housing and parks for economically disadvantaged or marginalized communities in other parts of the city.[13] Barcelona provides an alternative model, where smaller parks were more widely dispersed throughout the city as part of the capital improvements made in anticipation of the summer Olympics in 1992. This model might be relevant to the United States: Robert Garcia, Director of the Center for Law in the Public Interest, Los Angeles, and an advocate for urban parks, has made the argument that a web of small parks with more active recreation programs and facilities might better serve low-income and minority populations than more distant large parks.[14]

In addition, the natural and social ecologies of large parks are often at odds. The landscape architect Kristina Hill has argued that substantial ecological zones with extensive interior space—deep woods or wetlands, for instance—are crucial for certain species and an overall requirement in effective ecological function.[15] But these are precisely the spaces that are most problematic from a social perspective, being at once the most difficult to supervise and the least habitable for programmatic uses other than walking. Moreover, ecological function might be best where least disturbed by human presence. This is a matter not just of size but of configuration: social uses thrive around the edges and might best be accommodated in narrow, linear parks, while ecological functions demand width and depth. Can both ambitions be accommodated in the same place, or should recreational parks be separated from ecological reserves? The Bois de Boulogne is effectively managed as two units. Brigitte Seere, one of the two *ingénieurs forestiers* of the Bois, divides the park into two main categories: *massif forestier* and *horticole*.[16] The massif is as true a forest as possible, and not especially welcoming to people; the horticole consists of the cultivated lawns, lakes, and courts at the edges of the Bois, where the people are supposed to be. We are bound to see more of this division between social and ecological function and—perhaps—separate parks created for each purpose.

Perhaps most significant from a policy perspective is the question of how to arrest a pattern of physical and fiscal erosion of parks, especially in the United States. Physical erosion takes many forms, including the privatization of park facilities. But it also takes the form of land grabs: in the past generation, Griffith Park in Los Angeles has lost about 20 percent of its level area to freeway construction, which in turn has degraded the quality of adjacent land through air and noise pollution and severed the park from the Los Angeles River. Nearby Elysian Park lost hundreds of acres to a police training academy and Dodger Stadium. In Miami, half of Bayfront Park was privatized through a lease to the Rouse Company; the balance was turned over to a private trust that was required to cover most of its management costs through earnings from events, with the consequence that much of the park's remaining open space was transformed into fenced-in theaters used for fee-generating concerts.

Fiscal erosion has been just as damaging. Harnik's *Inside City Parks* provides compelling data for New York, which has one of the nation's oldest and best-documented parks departments. Between 1987 and 1996, inflation-adjusted public spending on parks in New York dropped 31 percent. The recreation program fared even worse: over the same period, its budget dropped 65 percent, from $20 million to $7 million.[17] Things improved a bit in the late 1990s, but after September 11, 2001, the city's budget picture darkened again. Moreover, parks are being milked for revenue, little of which goes back into maintenance or recreation. In 1998, for example, the city amassed $36 million from park department concessions, fees, and corporate promotional payments (it took in, for example, $700,000 from food concessions in Battery Park; $2.6 million from the operator of golf courses in city parks; and half a million from BMW for unveiling a new "James Bond"–model BMW in Central Park). A watchdog group called the Parks Council estimated that only five cents on every dollar generated by the parks department goes back to the parks; unfortunately, this is typical of municipal budgetary conduct across the country. At the same time, there has been considerable capital investment in park space in the city: a 6.5-mile stretch of riverfront on the west side of Manhattan, freed up by the collapse of the elevated West Side Highway, is being converted into the $360-million, 550-acre Hudson River Park in a notable public-private venture; about a third of it is already completed. But the occasional massive capital expenditure cannot by itself erase the more widespread physical and fiscal erosion of parks. Moreover, Hudson River Park will be a new-style landscape that depends on leases from moneymaking ventures within the park such as restaurants, health clubs, mini-golf, boat rental, and tennis courts.

This may be the most disturbing trend of all in contemporary parks: they are increasingly being expected to pay their own way. Parks are being designed not just for affordable upkeep but also for revenue-generating retail, recreation, and entertainment opportunities. The pastoral park is obsolete; parks are now looking more like commercial landscapes or entertainment destinations, with their intensive programs and numerous fee-generating activities. A little of this is probably harmless; it may even be helpful if a substantial portion of the proceeds can be captured for maintenance and recreation. In addition, revenue generation might provide a framework for combining social and ecological functions in the same large parcels: proceeds from programming might be used to offset the costs of preserving or creating significant ecosystems. But where does self-pay lead? The 1,480-acre Presidio in San Francisco, a decommissioned military base that is being converted into a national park, is required by authorizing legislation to be fiscally self-sustaining through rents and leases by 2013. This is the first time a national park has been designed around rent-paying residents, shops, and businesses, and the first case of a park agency (the Presidio Trust) being challenged to raise funds to cover all their costs—estimated at $37 million a year—or face having land sold off to private developers. This would be an instance of the ultimate privatization, and is perhaps an inevitable consequence of regressive social and economic policies that seek to keep public investment as low as possible while reducing taxes on the wealthiest citizens.

What recourse do we have in the face of such thinking? Parks already provide unaccounted economic benefits to cities: increased property values and thus enhanced revenues in surrounding neighborhoods, and advantages for their towns over other less green cities in the competition for tax-paying businesses and residents, to cite just two. We do not expect other public infrastructures—roads, water, sewers—to pay their own way, even if we do charge user fees for them. But user fees in the case of public parks compound the problem of unequal access to recreational and open space opportunities. Parks were once believed to be of value for their own sakes, as places of physical recreation and moral improvement. Large parks are now caught up in demographic, political, and economic transformations that they can neither hope to contain nor fully represent. The confident rhetoric of social engineering that rationalized the creation of large urban parks in the nineteenth century has been eclipsed—and has yet to be convincingly replaced. We might begin by affirming that unimpeded access to public parks is a crucial element of environmental justice. We need to reclaim parks as part of our essential urban infrastructure, as key features in functioning urban social systems. We need to maximize their potential as ecological systems, providing opportunities for climate moderation, stormwater infiltration, and biological diversity. We need to promote them neither simply nor primarily as revenue streams but as the vital laboratories of democracy—as the places where we go when we leave the protection of our families and neighborhoods and encounter the disparate, even contentious publics and counterpublics that compose the contemporary urban population. Parks are where we test the limits of our tolerance and our capacities for acknowledging difference. Less obvious but no less important, parks provide the necessary room for contest and struggle that allow people an opportunity to produce their own meanings and uses for public space. Whatever their flaws, parks remain among the most reliable places we have for the unscripted interactions that oil the creaky machinery of democratic social life.

NOTES

1. Katherine Anderson, "Large Parks: New Perspectives: Case Study: Bois de Boulogne," unpublished manuscript (2003), 11, 20.

2. Personal communication from Gina Ford, who studied Landschaftspark Duisburg-Nord for the 2003 exhibition and conference Large Parks: New Perspectives at the Harvard University Graduate School of Design.

3. Mario Schjetnan, conversation with the author, February 2006.

4. Data on Chapultepec Park is from personal communications with Mario Schjetnan, February 2006; it is drawn from an unpublished study of the park by the Program in Urban Studies at the National Autonomous University of Mexico, Mexico City.

5. Don Mitchell and Richard Van Deusen, "Downsview Park: Open Space or Public Space?" in Julia Czerniak, ed., *Downsview Park Toronto* (Munich and Cambridge, MA: Prestel and the Harvard University Graduate School of Design, 2001), 103.

6. Michel de Certeau, "Walking in the City," in *The Practice of Everyday Life* (Berkeley: University of California Press, 1988), 107.

7. Caroline Chen, "Large Parks: New Perspectives; Case Study: Tiergarten," unpublished manuscript (2003), n.p.

8. Personal communication from Rebekah Sturges, who studied Bos Park for Large Parks: New Perspectives.

9. Peter Harnik, *Inside City Parks* (Washington, D.C.: Urban Land Institute and Trust for Public Land, 2000), 20.

10. Information on Tijuca in this paragraph and next is drawn from Darren Sears, "Large Parks: New Perspectives; Case Study: O Parque Nacional de Tijuca," unpublished manuscript (2003), 11–13.

11. On Rio's program, see Rodolfo Machado, ed., *The Favela-Bairro Project: Jorge Mario Jáuregui Architects* (Cambridge, MA: Harvard University Graduate School of Design, 2003), and *Transforming Cities: Design in the Favelas of Rio de Janeiro* (London: Architectural Association, 2001).

12. Information on prostitution in the Bois is from Anderson, "Bois de Boulogne," 12, 22; in Madrid from a personal communication with Lara Rose, who studied Casa de Campo for Large Parks: New Perspectives.

13. Rosalyn Deutsche, "Uneven Development: Public Art in New York City," *October* 47 (1988): 3–52.

14. Robert Garcia, presentation, Large Parks: New Perspectives, April 2003.

15. Kristina Hill, "Urban Ecologies: Biodiversity and Urban Design," in Czerniak, ed., *Downsview Park Toronto*, 94.

16. Anderson, "Bois de Boulogne," 15.

17. Financial data for New York is summarized from Harnik, *Inside City Parks*, 14.

FIG. 1: James Corner/Field Operations, Singapore Gardens by the Bay, view of boat event at the mangrove, 2006.

JULIA CZERNIAK

Large Matters

Designing a "park for the twenty-first century" began at least by 1983, with the competition for La Villette in Paris. Almost twenty-five years later, this preoccupation continues in North America through such venues as Downsview Park in Toronto (640 acres, 1999), Fresh Kills Landfill in Staten Island, New York (2,200 acres, 2001), North Lincoln Park in Chicago (1,000 acres, 2003), and most recently, Orange County Great Park in the Los Angeles metropolitan area (1,000 acres, 2006). Each of these projects asks designers for forward-looking thought, prompting speculation on the roles for contemporary parks and, inevitably, their appearance. What is remarkable is that those of the last ten years are on very large metropolitan sites, suggesting an affiliation between largeness, visionary thinking, and the contemporary city. "Tree City" and "Lifescape," the winning schemes for Downsview Park and Fresh Kills, respectively, are now widely referenced by designers and others seeking strategies for contemporary large park design; and the last few years in particular have witnessed significant others internationally—such as James Corner/ Field Operations' scheme for Singapore Gardens by the Bay (2006) and Latz + Partner's Ayalon Park in Tel Aviv (2005)—and nationally, such as Hargreaves Associates' Orange County scheme (2005) and Clare Lyster's winning design for North Lincoln Park in Chicago (2003) (FIG. 1). Although varied in approach, these competition schemes share two characteristics essential to the vital social, ecological, and generative roles that large parks play in the contemporary city: legibility and resilience.

Legibility is a straightforward term, referring most commonly to the capability of something to be read or deciphered, such as handwriting. In our discipline, reading design work is more usefully framed as understanding its logics—textual, biological, organizational, and methodological. In the context of large parks, legibility is the capacity of a project to be understood in its intentions (its evolution and goals), identity (its distinguishing character and organization), and image (both its appearance, whether pastoral or post-industrial, and its marketing strategies). The concept of legibility extends from park design to the design process. In other words, to be realized, parks have to be legible to the people who pay for and use them. This legibility is a design challenge.

Resilience is a more complex quality. At its most general, resilience is the ability to recover from or adjust to change that may be perceived as "good" or "bad." Here, resilience is thought to be a positive attribute, both of character (as in a person) and of behavior (as in a material). As an ecological concept, resilience is the ability of a system to experience disturbance (such as a windthrow event, insect outbreak, or fire) and then return to a recognizable steady state.[1] In a less traditional and more useful ecological definition, resilience is "the ability of a system to adjust in the face of challenging conditions."[2] The important measure of an ecosystem's resilience is the amount of disturbance it can absorb while still maintaining its function before instabilities "flip" the system into another regime of behavior, such as grasslands becoming a shrub-desert.[3] As a tool for conceptualizing, planning, designing, and managing large parks, it is useful to think of resilience in this ecological sense. A park's capacity for resilience lies in the strategic design of its organizational systems and logics—whether infrastructure, form, or modes of operation—that enables it to absorb and facilitate change yet maintain its design sensibility.

However, given the complexity in living systems, as Nina-Marie Lister explains, "resilience as an ecological concept is not typically considered an empirical quality but a heuristic, useful to guide decision making that involves safe-to-fail trial-and-error methods and feedback through monitoring and plan or design evaluation."[4] In this way, a large park's ability to accommodate diverse and shifting social, cultural, technological, and political desires while maintaining its identity is a characteristic of its resilience. What matters in terms of a resilient park is the tension, in both design and management, between efficiency and persistence, constancy and change, and predictability and unpredictability.[5] Despite their differences, the park schemes I mentioned, and some iconic large parks that precede them, demonstrate that the scale, extent, and kind of disturbance that a landscape can absorb establish both challenges and opportunities for design.

The intentions of this chapter are threefold: first, to sketch the shifting relations of the park and the North American city over the last two centuries to highlight differences in their contexts and aspirations; second, to discuss the legibility and resilience of a selection of large urban parks, such as the winning schemes and current proposals for Fresh Kills Landfill, Downsview Park, and a few other recent projects;[6] and finally, to elaborate on the roles large parks can play in contemporary cities. Successful large urban parks are legible and resilient. The "park" is by no means obsolete, even though its relationship to the city, its design sensibility, and the preoccupations of those who shape them have changed dramatically.

Large Parks and the City: Shifting Relationships

To discuss large urban parks is by necessity to discuss the city. In her book on park design, Galen Cranz suggests that "whatever is decided about the function of parks will largely derive from some vision of the city, and it is by no means obvious how the city...is to be viewed."[7] How the city is viewed—as a formal and spatial organization, as an array of programs and events, as a

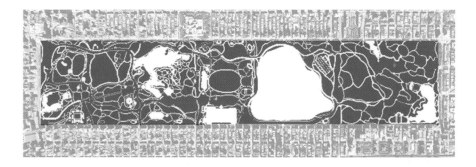

FIG. 2: Central Park as green object (top).
FIG. 3: Central Park's circulation strategy, as designed and as currently used (bottom).

set of problems and opportunities, as a metabolic organism, or as a political and administrative body—is crucial to understand the shifting relationships of parks to them.

In the half century leading up to the making of Central Park, the vision of New York City's future depicted in the Commissioner's Plan of 1811 is one of hills, streams, ponds, and swamps over which is superimposed a grid of streets and avenues. This plan allocated only 450 acres, constituted by seven squares and one parade ground, for parks. Fearing the health problems associated with such unrelieved density, civic leaders began lobbying for a large park.[8] Such space was readily available in the early stages of America's urbanization, when in many cases parks came before cities, and Central Park's subsequent expansion to 843 acres was seen as necessary to fulfill the park's primary role of offering relief from the industrial city and to provide the needed space to layout picturesque geometry. The details of "Greensward," Frederick Law Olmsted and Calvert Vaux's 1858 winning competition scheme for this land, are well known; revisiting them here is a response to their frequent mischaracterization. Regarding Olmsted's philosophy simply as escapist and the nineteenth-century park as separate from the city sets up a false polemic—problematic because of the successful strategies Central Park employs that are still relevant to making large parks today. Although this park was designed to produce the effects of a rural landscape to offer relief from harsh urban conditions, the park/city relations Olmsted and Vaux envisioned are more complex than this simple opposition might suggest. As landscape historian and critic Elizabeth K. Meyer points out, "It was a place of nature, but it was not, by any means, anti-urban."[9] A few examples illustrate this. First, Central Park is *both* a respite from the city—a place where, in the "absence of distraction," nature could produce restorative effects—*and* a spur to urban development.[10] Olmsted understood the relationship between building parks and building cities—that the initial cost of land would be more than compensated for by inevitable development of the park's perimeter. Second, the park is *both* an artificial place made through considerable construction processes *and* an image of nature, where city functions and facts were screened with lush planting. Its appearance mattered. The park's figural void—strongly legible—contributes to its reception as a green object. Long after its social, recreational, and infrastructural agendas fade from memory, the image of the park lingers, carrying with it the park's primary if not overgeneralized role (FIG. 2). Finally, the park accommodates *both* the functional demands of the future city (such as traffic, through sunken transverse roads) *and* the formal requirements of a picturesque landscape, seen in the organizations of internal roads and pedestrian paths. The first strategy extends the grid, the second resists it.

Olmsted's brilliant depression of the transverse roads and his separation of circulation paths in plan and section enable the park's resilience. Despite 150 years of programming and development pressures that have transformed many of the park's green spaces (the disturbance), the key sensibility of the park, the ability for a visitor to flow seamlessly—on foot, on a bike, in a car, or on a horse—without interruption, persists (FIG. 3). Central Park, like any living system, is not monolithic and static either ecologically or culturally. The original infrastructural armature for the park enables the addition of buildings (such as the Metropolitan Museum of Art); the conversion of space (from the reservoir to the Great Lawn to ball fields); and the shift of management strategies (that allow the Ramble to be a place for either people or birds) without the park's essential character being destroyed.

The urban context for parks in North America has changed significantly since Olmsted's time. In *Postmetropolis*, urban critic Edward Soja speculates on the new patterning of urban form that has emerged as a result of globalization and economic restructuring.[11] He offers an array of terms to describe these reconfigured spatial conditions—Megacity, Outer City, and Edge City— each with landscape implications. Megacities, or the "enormous population size of the world's largest urban agglomerations," characterized by "polycentric, almost kaleidoscopic socio-spatial structures," produce vast urban landscapes where the traditional defining quality of the city, density, no longer pertains (FIG. 4).[12] Outer City "arises from a process involving the *urbanization of the suburbs*," with its own spatial condition and lifestyle revolving around the car, the single-family house, and the private garden, giving rise to the privatization of landscape space.[13] Edge City implies "buildings that… dot the landscape like mushrooms, separated by greenswards and parking lots"—a form contingent on malls and office space, which produce multiple cores.[14] These terms all describe a loss of the dominance of the metropolis's distinctive organizational form, which implies a radical rethinking of large urban parks.

That landscape appears as an element of these urban formations is widely recognized, especially with the rise of subdisciplines of architecture and landscape such as landscape urbanism. In the most cited example, Dutch architect Rem Koolhaas uses "landscape" as a descriptor for all of Atlanta, suggesting that density is here replaced with "a sparse, thin carpet of habitation…. Its strongest contextual givens are vegetal and infrastructural: forests and roads (FIG. 5)."[15] However, parks per se, with their association of social programs, civic amenities, and perhaps ecological sensitivity, account for less than 4 percent of downtown space within this presumed landscape surplus. This description raises a question: in a dispersed metropolis where landscape is everywhere, is the urban park at risk? As graphic designer Bruce Mau—one of the designers for Downsview Park—suggests, "to imagine a park presumes an urban condition. When Frederick Law Olmsted imagined Central Park, he imagined the context it would eventually inspire and sustain."[16] Imagining the context of the contemporary metropolis is crucial for park designers.

Consider the sites for nineteenth-century and twenty-first-century urban parks in North America as described by historian Sam Bass Warner when projecting future roles for parks. Whereas the industrial city was characterized by overcrowding, disease, and long work days, the contemporary city is distinguished by "a superabundance of public and private open spaces, not overcrowding of the land; national and global environmental problems, not only local ones; and…a commercially managed fantasy life, not an imaginative [one]."[17] Warner further suggests that the current urban context presents spatial and social disconnection, fragmented authority, the rise of individual interests, and a lack of identity. Quite a set of challenges.

One obvious implication of these different urban contexts is the configuration of new park sites—their shape, perimeter, and interiority. Parks that emerged in relation to or even before the city, such as Central Park and Golden Gate Park, have strong and easily remembered figural forms. With the irregular parcels of land available in today's city, such as reclaimed industrial wastelands and decommissioned airports, a park's configuration is imposed more than chosen. This is problematized further in projects whose designs expand beyond their given lots, such as Bernard Tschumi's Downsview Park scheme, which proposed linkages beyond the competition boundaries to adjacent river ravines, making the domain of the park even less legible. So in addition to questions of a park's legibility that stem from recognizing its limits—"where is the park?"—large park schemes with unconventional configurations provoke other uncertainties—"how does it look?" and "what can it do?" These questions require designers to both reconcile the park's visual relationships to its surroundings and foreground how parks work.

The contemporary urban mosaics of which these large tracts of land are a part are often fragmented socially and programmatically, posing challenges from an ecological standpoint regarding networks, corridors, and connectivity—key components of a landscape's ability to maintain biodiversity. The sites themselves are degraded and usually in need of remediation. These irregular configurations, however, are actually assets. Proposed parks with shapes like the one that runs through the redeveloped Stapleton Airport in Denver resonate with landscape ecologist Richard Forman's sketch for an optimal patch configuration, helping to facilitate species distribution (FIG. 6).[18]

Contemporary large park sites also raise questions of density. For example, in Toronto, where private open space is abundant and population is dispersed rather than overcrowded, who are these parks for? The site for Downsview Park makes sense for a large urban park only in the context of the Greater Toronto Area's anticipated growth by more than two million people over the next twenty years. Robert Glover, former Director of Urban Design for the City of Toronto, describes the parklands as "located in the midst of one of the city's major potential suburban intensification areas," and asserts that the park will play an integral part in the city's attempt to intensify itself.[19] One of the initial goals of the park, very different from parks of the past, is to help produce density, not provide relief from it.

Large parks in the contemporary city also have to contend with the complex and often contentious histories of their sites. Whereas Olmsted and Vaux

FIG. 4: North America at night showing large urban agglomerations (top).
FIG. 5: View of landscape surplus in North American cities such as Atlanta (bottom).

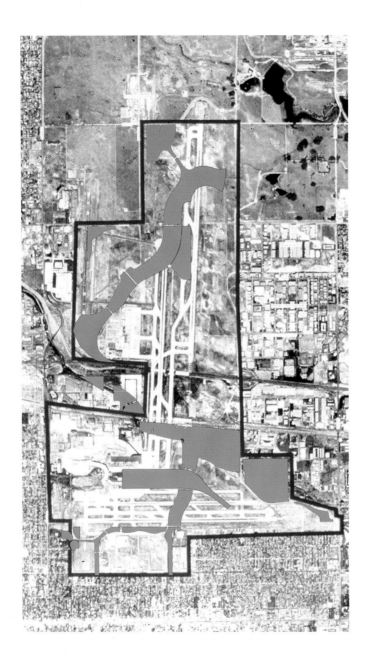

FIG. 6: Heather Vasquez and Nicholas Rigos, diagram of proposed park system running through the redevelopment of the Stapleton Airport site, Denver.

modified rock outcroppings and swamps to make Central Park, contemporary sites such as that for a large park at Fresh Kills or Downsview have already been transformed from wetlands to landfill, or from forest to field to air-force base, respectively. In the face of such layered complexity, designers must decide which conditions to address and which to ignore. Moreover, these layers are not neutral, freighted as they are with social and political implications. For example, the design for Landschaftspark Duisburg-Nord in Germany shies from any reference to the site's affiliation with the Nazi party during World War II.[20]

Finally, although economic and political challenges are a part of any design process, regardless of their time, the context for these challenges shifts. Although parks were planned by the apparently homogeneous elite in the nineteenth century, there were nonetheless tensions among individuals, and Olmsted was fired several times by New York City politicians during the evolution of Central Park. Today, large park design almost always happens with a participating public, and the schemes undergo an extraordinary amount of public and regulatory review that gives the planning and design process social, cultural, technical, and ecological input. The schemes most resilient to feedback are the most likely to survive.

New York City/Fresh Kills Lifescape

New York City is an apt context in which to examine how parks might address these contemporary challenges. When Central Park was under construction, the core of New York City's population was 2.5 miles south of the site. Now the park is thoroughly absorbed in the New York metropolitan area, an agglomeration that spreads across a thirty-one county region, engages three states, and is populated by more than twenty-one million people. That the country's largest landfill is also located here is not surprising; what is surprising is the continued progress being made in transforming the Fresh Kills Landfill into "Lifescape," New York City's largest park.

Lifescape is the winning scheme for the 2001 Landfill to Landscape competition, a two-stage international competition for a model land-regeneration project. Led by James Corner/Field Operations, the Lifescape project team successfully addresses a set of urban problems that the competition brief reformulated as goals. Socially and ecologically, Lifescape creates connections at nested scales, from the local site to the region, providing for flows of people, water, and wildlife, as well as recreation and educational opportunities. Technologically, it complies with all landfill regulations while creating phased public use that responds to significant landfill processes, not merely time. Aesthetically, the scheme reveals the unique character of a site that performs as a land-regeneration project and, despite its appearance as a grassland, is thoroughly urban in its synthetic nature.

The conceptual view of the city that drives both the competition scheme (2001) and its subsequent development into a draft master plan (2006) is not one that has been conditioned through the nineteenth-century lens of difference and opposition, where parks such as Central Park (with its "softly undulating, virtuous and pastoral nature") are set up as what Corner calls "great vessels that provide respite from the negative effects of urbanization."[21] Instead, the city is thought of as a metabolic and technologic machine, what Sanford Kwinter calls a "soft system"—adaptive, flexible, and evolving through its capacity to absorb and exchange information with its surroundings.[22]

Lifescape positions itself as part of this dynamic set of interrelations, where the park is an analog to the city's spatial and temporal interactions. What the Corner team calls its "nature-sprawl" sponsors a compelling view of nature that they explain is "no longer the image we look at, but the field we inhabit."[23] Their technological view of nature as a "bacterial machine" promises to evolve "new life-form opportunities" on Fresh Kills' sterile terrain that "makes use of the logics of natural systems and the dynamics of ecological feedback without the romantic attachment to a pastoral idea of nature."[24] The scheme is an "infrastructural strategy of emergent colonization that stages various systems and sets in motion a diverse ecology of events and the complex organizations of forms."[25] The proposal thoroughly outlines a successional strategy to establish a landscape ecology by remediating the soil, stabilizing the slope, removing invasives, establishing grassland, introducing woody material, and adaptively managing. Over time habitat types, with many native species, will emerge across Fresh Kills, responding to the specific characteristics of their local sites. For instance, the moist and protected north slope supports mixed deciduous woodlands, while the dry exposed south slopes are planted with drought-resistant prairie grasses and pine/oak barren islands. The result is a landscape that promises ecological and social interconnectivity at many scales, defragmenting an urban mosaic as it extends beyond the boundaries of the site. Lifescape envisions landscape as a catalyst to transform the whole island, weaving together its disparate parts. Lifescape's agency promises to free Staten Island from the shadow of Manhattan, making it a more vital metropolis of its own.

As a landscape proposal, Lifescape is believable in its goal to recast Staten Island as its own strong place, no longer reliant on the historic city center. Yet with polemics that so pointedly resist nature/city oppositions, its competition "image" is unapologetically natural-looking, not that different from the soft and undulating nature of which Corner was critical (FIG. 7). Although circulation plans, habitat restoration, defragmentation strategies, and prominent external views render Lifescape seamless with the city, the project could be understood as a restoration effort with a management strategy, shored up through the choice to neither index the landfill nor reveal the landscape's own artificiality, strategies that Corner suggests have lost their critical power because of the current tendency to view nature as wholly synthetic.[26]

Admittedly, however, when one looks more closely at the plan details of the winning entry—the forest threads that slip off the slope to wrap significant

FIG. 7: James Corner/Field Operations et al., "Lifescape," Fresh Kills competition proposal, aerial view of park.

spaces, the combed earthwork that supports planting, or the circular clump forests in the siltation basins—each suggests a sort of plasticity and constructedness that is not simply a naturalism (FIG. 8). Field Operations is less concerned with their park's image (how it looks) than they are with its identity (its distinguishing character), evident in its legibility and resilience. In their competition entry, the team recognized the park's role in repairing ecological damage and bringing a more diverse and self-sustaining mix of species, programs, and processes to the site. They also acknowledged another task: to create a constituency that will maintain and nurture the landscape over the long term. To achieve this, Corner suggested that the landscape must be legible to its users, a condition less evident in a perspective drawing than in the "cultivated understanding of its evolutionary logic."[27] Their planting strategy promises to enable residents and visitors to understand the condition associated with its varied planting communities, a desire made evident to the competition jury in the extensive diagrams and drawings in a book that accompanied the scheme.[28] The scheme's legibility extended to its plan map figure—referred to as "part fetal form, part flower, part map, part open-ended"—an easily recognized identity tool for promoting Lifescape to the public during the planning process (FIG. 9).[29]

As a design process and a draft master plan, Lifescape delivers on these promises. Field Operations' community outreach program has been ambitious in its scale: designing advertisements for the city's website, widely disseminated posters for public meetings, transformable project logos, and graphics for billboards, bus ads, and even the side of Department of Sanitation trucks (FIGS. 10–12). These measures are intended to generate a high degree of public interest and participation in the master-planning process and stand for a park in planning, one yet to exist, in recognition of the crucial impact that the legibility of the park and its operations will have on its future life.

Lifescape's resilience is the second key component of its identity. The view of nature it embodies (a working system rather than a view) will enable the park, when built, to absorb and react to ecological feedback not under a designer's control. The Corner team also understands that one of the more significant characteristics of the city is its role as a governing authority that will, in the best of circumstances, accumulate capital, ensure environmental protection, and engage the public to enable the project's realization. The challenges of engaging the city are immense. Since winning the competition, the team has been through an elaborate planning and design process and has been resilient enough to respond to the needs and desires of the community as circumstances have dictated. The current master plan demonstrates Lifescape's reslience in that the key sensibility of the scheme, the capacity to handle and process movement and change through its organizational logic, is still present. As such, Lifescape can be seen as a textural field—a pixel map— where effects are projected and managed, added to and erased while retaining overall legibility (FIG. 13).[30] In the current draft master plan, the conceptually

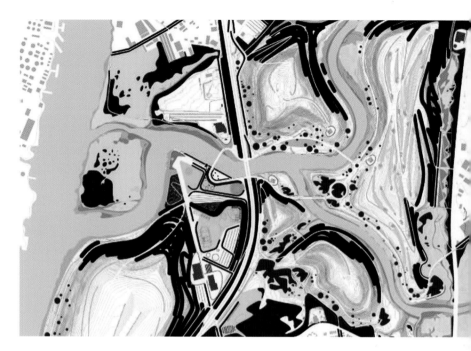

FIGS. 8, 9: James Corner/Field Operations et al., "Lifescape," Fresh Kills competition proposal, competition plan detail (top) and plan map figure (bottom).

87% OF AMERICANS
ENJOY 5 OR MORE
OUTDOOR ACTIVITIES.
WHAT ARE YOURS?

FRESH KILLS IS
BECOMING A PARK.
HOW? COME SEE:

MARCH 24, 7-9PM
PUBLIC MEETING AT
HOLY TRINITY CHURCH
1475 RICHMOND AVE.

NEW YORK'S
NEW PARKLAND
FRESH KILLS · lifescape

KILL IS A WORD OF
DUTCH DERIVATION.
IT MEANS A STREAM
OR A BROOK.

NEW YORK'S
NEW PARKLAND
FRESH KILLS · lifescape

NEW YORK'S
NEW PARKLAND
FRESH KILLS · lifescape

lifescape

FIGS. 10–12: James Corner/Field Operations, "Lifescape's" identity, community outreach, and communication design strategies, 2006 (above).
FIGS. 13, 14: "Lifescape" as pixelated field (opposite, top); and draft master plan, 2006 (opposite, bottom).

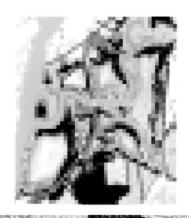

driven spatial systems of "threads, islands, and mats" have evolved into the more conventional terms of "program, habitat, and circulation," having succeeded at evolving the design. Furthermore, the park can now be understood by the significant places that its organizational logic enabled: the Confluence and the North, East, South, and West parks. In this way, the scheme's current naturalism is well designed, responsive as it has been to cultural, political, and technical feedback (FIG. 14).

Furthermore, Lifescape advances and clarifies important distinctions between two sets of terms often used interchangeably in the design disciplines. A resilient scheme such as Lifescape presents a strong design logic that can sustain a dialog with multiple contexts, accommodating and growing from pressures put upon it. This is a different concept from "emergent" (where the design cannot be totally predicted) or "open-ended"(where multiple possibilities coexist). Given the ubiquity of landscape projects that use these terms interchangeably, this distinction should not be underestimated. The threads, islands, and mats of the original competition scheme have evolved as needs are redefined, yet enable the scheme's overall robust identity to remain, not unlike the infrastructure of Central Park.

Finally, Lifescape makes a key distinction between "adaptive" and "resilient" as design concepts.[31] Whereas adaptation suggests continual change in form and identity to adjust to a set of conditions (from field to meadow to forest), resilience implies a return to a recognizable state after a disturbance. Whereas Lifescape's phasing strategy promises a "strong initial landscape organization in the early years sufficiently resilient to absorb any number of future permutations," it plans for adaptation within its framework.[32] This distinction leads to a powerful strategy for parks—a landscape type that simply cannot be adaptive in its entirety because of the need to be recognized, used, and invested in.

Toronto/Downsview Park's Tree City

The strategies for Downsview Park, the Canadian government's first national urban park, reiterate and expand on the concepts of legibility and resilience. The morphology of Toronto can be understood as successive generations of suburban development radiating outward from its historical center. When the air force base was established in 1947, the future park site was on the edge of the city. Today it is in the middle of the emerging post-metropolitan condition known as the Greater Toronto Area, an urbanized territory with a population of 4.4 million. The Downsview Lands are a large redevelopment site within this urban expanse, a 640-acre urban void edged by aging postwar suburban housing and industry, second in size only to the land intended for Lake Ontario Park, a 925-acre site currently being designed. The park competition site includes 320 acres, with the remaining land reserved for revenue-generating development (FIG. 15).

FIG. 15: Downsview Lands in the context of Toronto. The future Lake Ontario Park site is to the lower right of the image.

That the competition was based on seemingly antithetical goals is no small matter. The parklands are centrally located in the midst of one of the city's "suburban intensification areas," and the park is to play an integral role in the city's attempt to intensify. Infrastructural facts that will aid in these efforts—the airstrip, the highway, and most important the proposed transportation hub—are in the park. But at the same time, the park is to be a model of ecological and economic sustainability. The conurbation of which the site is a part corresponds to the Greater Toronto bioregion, a basin formed by the Niagara escarpment, the Oak Ridge moraine, and Lake Ontario (FIG. 16). In this context, and amid the growing desire to conceptualize development within the region's ecosystems, the park presents a major opportunity to contribute to the city's system of green spaces. The site is on a topographic high point between the Don and Humber rivers, two of the larger regional watersheds. Its adjacency to ravines, characterized by degraded water and urban use, suggests the possibility for connecting and enhancing flows of water, habitat, and wildlife. These ravines, a strong component of Toronto's urban identity, are spines for much of the city's parkland, with a mix of recreational, institutional, and commercial functions (FIG. 17).

"Tree City," the winning competition scheme by a team led by graphic designer Bruce Mau, is now well known as a diagram for growing a park into the city. Positioned within Toronto's tradition of meager expenditures on green space, Tree City promised to become a green destination for relaxation and recreation, an infrastructure that increases value over time, an instigator to suburban densification—in short, a civic amenity.

Tree City was advocated as more of a formula than a design—a pragmatic response to unknowable political and economic conditions. Its polemic goes: grow the park + manufacture nature + curate culture + 1,000 pathways + destination and dispersal + sacrifice and save = low-density metropolitan life.[33] The formula suggests the park's emergence at two scales. At the site scale, the scheme begins with soil preparation that will support planting over time. The park will eventually be 25 percent forest, in addition to meadows, playing fields, and gardens, with 1,000 paths for varied uses. At the scale of the city in a later phase, the scheme's landscape clusters and pathways promise extension into adjacent areas, weaving into the suburban fabric to link up with Black Creek and West Don River systems and ravines. Tree City's organizational strategies sprawl across the site and invert typical figure/ground relations, promoting landscape's legibility as figural density against a less dense suburban ground. The signature clusters, or dots, that appear in the scheme and its representation are how Tree City is popularly recognized (FIG. 18).

Almost seven years after winning the competition, that Tree City has evolved from a polemic to a preliminary plan for land that has still to be completely acquired from the Defense Department is remarkable—an index of the slow progress, ambiguous vision, and sometimes contentious process of shaping the park. Tree City's proposed formula hovers between fantasy and

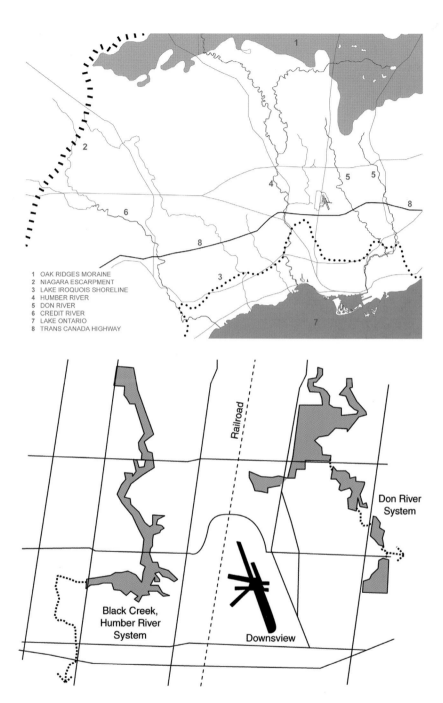

1 OAK RIDGES MORAINE
2 NIAGARA ESCARPMENT
3 LAKE IROQUOIS SHORELINE
4 HUMBER RIVER
5 DON RIVER
6 CREDIT RIVER
7 LAKE ONTARIO
8 TRANS CANADA HIGHWAY

Railroad

Don River
System

Black Creek,
Humber River
System

Downsview

FIG. 16: Bernard Tschumi Architects, et al., Downsview Lands in the context of the Greater Toronto bioregion (top).
FIG. 17: Kristina Hill, Downsview Park relative to the Don and Humber river ravines (bottom).

FIG. 18: OMA + Bruce Mau et al., "Tree City," Downsview Park, Toronto, winning competition scheme.

folly, and one wonders how much of the park's current status is a result of the open-endedness of a design as formula. As a "pragmatic response to unknowable conditions," the scheme is difficult to sell by politicians and planners that have a hard time knowing for what they are rallying. A recent article in the *Toronto Star*, "If Downsview Park Matters, Why Has Nothing Begun?" reflects the public's frustration, arguing that after seven years and not much action there is cynicism about its future.[34] Yet Tree City can still be seen as "performing." An interim conceptual master plan unveiled by Mau and his team in February 2004 vaunted the theme of "six hundred acres of ecologic, economic, and social sustainability."[35] A comprehensive park plan was completed in July 2006, and a modest test plot of 30 acres of urban reforestation is currently under way (FIG. 19). A QuickTime video is posted on the park's website that supplements ambitious descriptions of its future with specific images (FIG. 20). However slowly, each iteration fills in the gaps of the previous one.

Yet the integrity of the eventual plan for Downsview Park remains to be seen. If the diagram of Tree City works, the conceptual master plan should be seen as nothing but a "plan du jour"—a moment in its intended evolution. One can imagine that the specific configuration within the park's three zones—the action zone to the north, the forested promenade in the center, and the cultivation campus in the south—can be easily changed without the park losing its identity. Seen this way, the future Tree City and the competition formula that enables and promotes it are legible and resilient in significant ways.

Shortly after the competition, the clusters, or dots, that appear in the scheme and its representation generated much speculation. In one way, the dots allow the park to be understood in its intentions, organizations, and imagery—it is legible. The circle pattern makes implementation easy, in that whatever goes into the park takes its form. This move accommodates conflicting desires while staying coherent. In addition to the scheme, the idea of legibility was intended to permeate all levels of the work, including maintenance and the public process. Just as important, their big green dot, understood as the park's logo, gave it an immediate and marketable identity. Bruce Mau has explained in many contexts how the graphic design of the project helped to illustrate its advantages, and he advocates using commercial principles and practices to do so.[36] In the years following the competition, the green dot has proved successful as an identity tool. The intention was that once the park started to evolve, this operational image would be supplanted by actual ones, and it has been. The new logo for Downsview Park, a rotating cluster of bodies/buds, is derived from the green dot (FIG. 21). The field of dots are now replaced by formations with material specificity. Tree City's legible strategies have turned environmentalism into a form of commercialism, engaging a mainstream mentality that should propel the project.

In Tree City, resilience is linked to legibility. The circle pattern that gives the park its identity is also resilient in the interchangeability of the

CHESSWOOD
NEIGHBOURHOOD

WILLIAM BAKER
NEIGHBOURHOOD

SHEPPARD
NEIGHBOURHOOD

CULTURAL
COMMONS

DEPARTMENT OF
NATIONAL DEFENCE

ACTION
ZONE

ACTION SPORTS
COMPLEX

ALLEN
NEIGHBOURHOOD

PROMENADE

BOMBARDIER
LANDS

TTC
WILSON YARDS

CULTIVATION
CAMPUS

STANLEY GREEN
NEIGHBOURHOOD

SHEPPARD AVE WEST

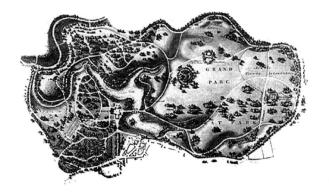

FIG. 19: Downsview Park: Sustainable Community Development Plan, October 2006 (opposite, top).
FIG. 20: Stills from video, "Downsview Park: The Future: A Virtual Tour" (opposite, bottom).
FIG. 21: OMA + Bruce Mau et al., "Tree City," Downsview Park, winning competition scheme, logo; Parc Downsview Park, logo.
FIGS. 22–24: Blenheim, England, views of tree clusters (middle, left and right) and plan (bottom).

circle's spatial location and material specificity (tree massing to water pool to building) to, most important, the capacity of parts of the park system to stay closed to disturbance (which stabilizes the design) and others to be continually affected by it. In her discussion of Tree City in *Downsview Park Toronto*, design critic Anita Berrizbeitia advances ways in which landscape design projects can address environmental and ecological scales while still articulating issues of meaning, artistic expression, and language by their various components being either "open" or "closed" to external perturbations.[37] Using Berrizbeitia's framework, we can think of the dots as closed to the disturbances that the rest of the park, as it evolves, is open to. This is not new, or even surprising, just a little repressed. For example, think of the tree clusters in the eighteenth-century landscape garden Blenheim (FIGS. 22–24). One can imagine the topographic field conditions as a way to guide stormwater, nurture native plant communities, and encourage species diversity, while the tree clusters remain neatly trimmed and maintained in circles.[38] This distinction between elements that are static and elements that play provisional roles is essential to a park's resilience.

Large Parks Now

Central Park, Lifescape, and Tree City sponsor a conceptual framework of legibility and resilience. Three recent projects for large parks take the concepts further and, in turn, start to prompt questions. "Assembled Ecologies: Infrastructure à la Carte," a winning scheme for North Lincoln Park in Chicago by a team led by Clare Lyster, Julie Flohr, and Cecilia Benites, proposes an organizational system based on the dimensions of a Chicago block. The organization of the park through an infill mechanism of five morphologies of "modular infrastructural tiles" enables resilience in a way that lets various stakeholders make an individual statement on their tiles without compromising the vision for the park. This resilience is political, recognizing the complex and competing interests of various constituencies and "enabling the production of contemporary public space."[39] Like Tree City, the organization system that enables the scheme's resilience is optically legible, in that everything that goes into the park takes its form (FIGS. 25–28).

In James Corner/Field Operations' scheme for the Singapore Gardens by the Bay project (500 acres), resilience is enabled by the organizational structure of the overlaid circular geometries, creating a large rolling topographic surface within which various programs can be nested. However, the organizational *logic* of porosity and flow at numerous scales is the innovation here. The various materials allocated to the different surfaces allows them to perform in very different ways, thereby lending an overall resilience to the field as a whole but with maximum variation and flexibility within the parts.[40] Glass, turf, concrete, and stone, used in various combinations, suggest infinite exchangeability based on user wishes, cost, sitespecificity, or lifespan. In other words, where materials perform, not signify, the only criteria is porosity. Moreover,

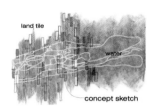

land tile

water

concept sketch

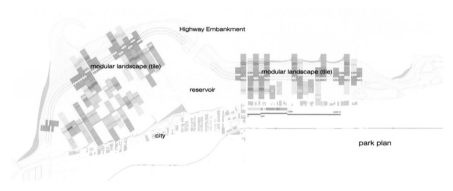

Highway Embankment

modular landscape (tile)

modular landscape (tile)

reservoir

city

park plan

FIGS. 25, 26: Clare Lyster, Julie Flohr, and Cecilia Benites, "Assembled Ecologies," North Lincoln
Park, Chicago, concept sketch and park plan (top); and view of model (bottom).

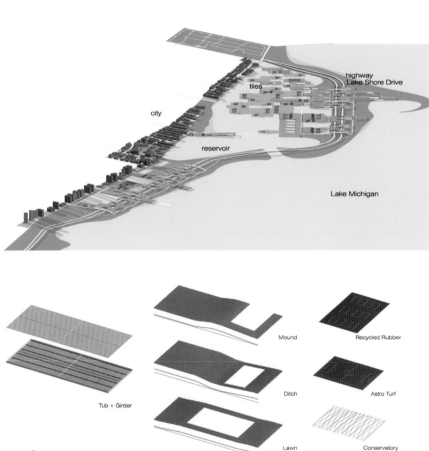

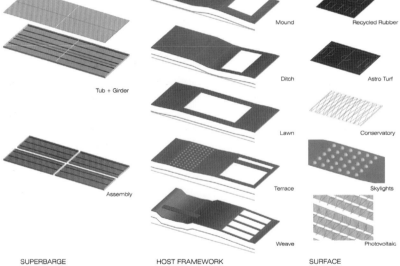

FIGS. 27, 28: Clare Lyster, Julie Flohr, and Cecilia Benites, "Assembled Ecologies," North Lincoln Park, Chicago, axonometric (top) and tile taxonomy (bottom).

stormwater is processed within the various pores of the surface, flowing from one pore to the next in a successive sequence of appearance and disappearance. Finally, the reforestation strategy for the park is based on the same organizational geometries, layering species to create a rich textural canopy (FIGS. 29–31).

The legibility of Hargreaves Associates and Morphosis's scheme for Orange County Great Park is based on an understanding of its site-specific logic. Its resilience resides in a middle scale of organizational logic that addresses the large, complex nature of its 1,000-acre site. Each of their five driving "constituencies" (water, nature, activity, culture, infrastructure, sustainability), although not organizational systems themselves, rely on various logics to govern their formation and evolution. For example, "nature" is constituted by landscape types such as grasslands, irrigated fields, riparian systems, and productive plots; "activity" is conceived as subsystems of patches (irrigated pads with embedded services) and strips (that accommodate sports and recreation); and "infrastructure" is divided into the fixed and the flexible, recognizing the difference between aspects of the park instrumental to its long-term function and that which is more open to changed configurations, based on user wishes, economics, politics, etc., such as the circulation networks of bicyclists and pedestrians.[41] Conceiving of the park this way, even when not evident in its plan arrangement, is a strategy of resilience (FIGS. 32–36).

Legible and resilient parks directly and provocatively engage the debate that has emerged in landscape discourse that tends to position landscapes as either overly designed and apparently static (based on how things look and what they mean) or underdesigned and infinitely malleable (those that foreground, sometimes as organizational strategies and sometimes as polemics, landscape's emergent properties). All landscapes fit somewhere in between. Perhaps it is useful then to locate the design of large parks along a spectrum, with the more resilient to one side and the less resilient to the other.

Probing this idea further, however, leaves questions that invite speculation: What is the relationship between legibility and resilience? Is the complexity of large sites prompting these strategies? Can parks be resilient at small scales? Is there a link between provisional form and resilient concepts? How else might these concepts be developed beyond new organizational strategies and logics that challenge the fixedness of composition?

Large Parks: Projecting Futures

Large parks can play vital roles in the city in three ways, and their legibility and resilience are measures of their ability to do so. First, as *social catalysts*, the schemes I've discussed promise contact and exchange for people in otherwise disjointed urban environments through an array of spaces, activities, and circulation systems. In the extensive lists of invariably conflicting programs that accompany each proposal, one can find amenities such as daycare centers and community gardens, places that could enlarge what Cranz cites as the reduction of the range of social functions performed by parks.[42] Furthermore,

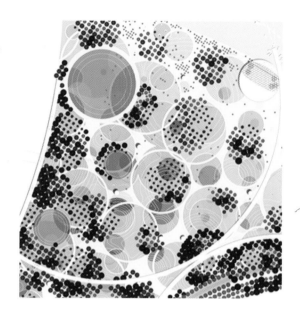

FIGS. 29–31: James Corner/
Field Operations, Singapore
Gardens by the Bay, 2006, site
plan (opposite); plan detail
(top); and view of scales of
porosity (bottom).

the processes by which these parks emerge imply a social dimension, where the public grows and learns through the design process. Environmental interests (say, the presence of wildlife corridors) have to be balanced with often conflicting cultural desires (such as for activity spaces). The interactive process of exchange, feedback, and growth between invested constituencies (and their understanding of the scheme) as well as a design's ability to accommodate these conflicting wishes is a measure of its legibility and resilience.

Second, as *ecological agents*, these parks promise, in diverse ways, to act more like what Alan Tate has coined a "heart" than a "lung,"[43] enabling life through ecological investment locally and across the larger horizontal urban field. By their size and configurations, these schemes facilitate interconnectivity across a fragmented landscape, responding to the obvious threat that urban expansion poses to landscape vitality and biodiversity. These parks are places where designers can develop constituencies that will understand, nurture, and maintain them—a positive collective act—because the aspirations of *their* large park in *their* contemporary city are legible to them. Furthermore, these parks work ecologically, not just look like they do. Resilience here is their literal capacity to recover and adjust from storms, pests, disease, fire, and other natural disturbances.

Last, parks continue to be *imaginative enterprises*. Cranz suggests that parks keep "on hold" values threatened by the facts of city life until culture can reincorporate them, by showing the difference between itself (the park) and ordinary reality (the world). Parks can be "places for imagination to extend new relationships and sets of possibility."[44] Here, parks are places to project futures, and resilience is a measure of their ability to hold competing ones. As such, how these parks appear matter. They must look different from their surrounding contexts or they, and their agendas, are simply are not recognizable. Legibility is a part of this identity building.

The large, the park, the city, and the future are intimately related. We (those who design, use, and enable parks) need to locate the specific potential of park design discourse, where concepts like emergent, open-ended, adaptive, and resilient are readily understood. We also need to recognize that parks are something that, both literally and metaphorically, must be cultivated. Those within and outside the design disciplines need to understand why parks are necessary, the roles they can play, and how they can look. Designers also must stop the purely polemical call for the end of parks as a landscape form and to instead embrace them, knowing that the nineteenth-century definition of the word is outmoded. Legibility and resilience offer strong starting concepts for this project.

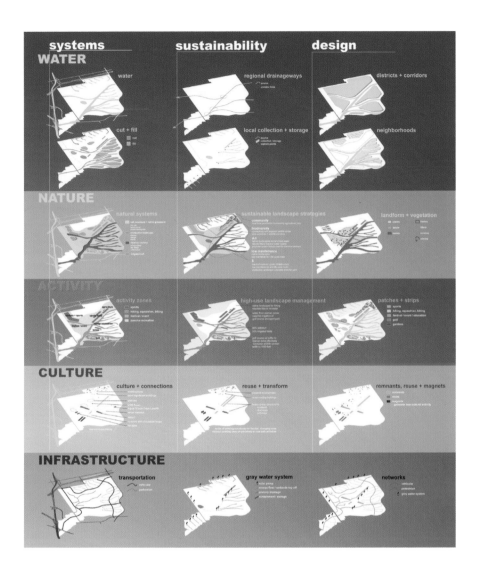

FIGS. 32, 33: Hargreaves Associates et al., Orange County Great Park proposal, diagram of five "constituencies" (above) and phasing (overleaf).

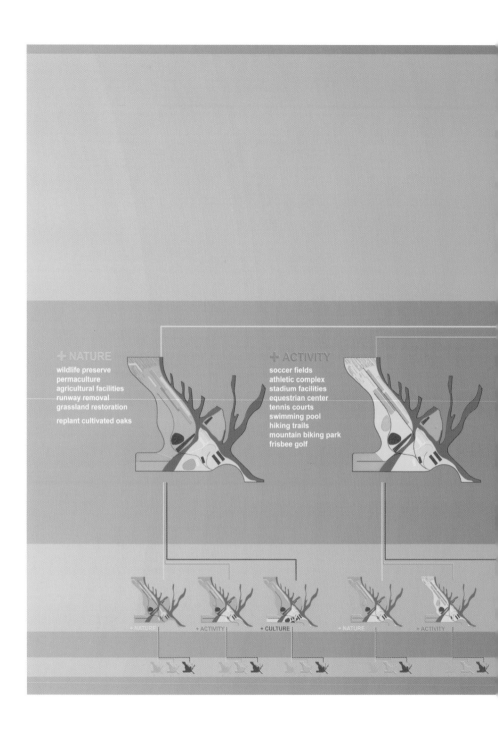

+ NATURE

wildlife preserve
permaculture
agricultural facilities
runway removal
grassland restoration

replant cultivated oaks

+ ACTIVITY

soccer fields
athletic complex
stadium facilities
equestrian center
tennis courts
swimming pool
hiking trails
mountain biking park
frisbee golf

+ NATURE   + ACTIVITY   + CULTURE   + NATURE   + ACTIVITY

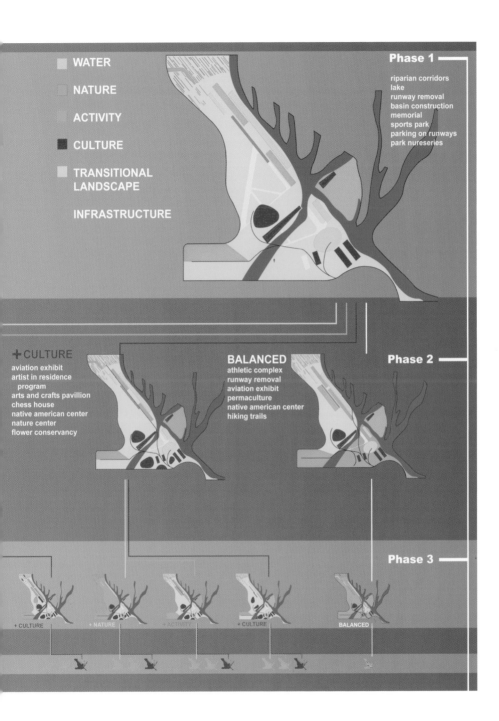

WATER

NATURE

ACTIVITY

CULTURE

TRANSITIONAL
LANDSCAPE

INFRASTRUCTURE

Phase 1

riparian corridors
lake
runway removal
basin construction
memorial
sports park
parking on runways
park nureseries

+CULTURE

aviation exhibit
artist in residence
   program
arts and crafts pavillion
chess house
native american center
nature center
flower conservancy

BALANCED

athletic complex
runway removal
aviation exhibit
permaculture
native american center
hiking trails

Phase 2

Phase 3

+ CULTURE   + NATURE   + ACTIVITY   + CULTURE   BALANCED

FIGS. 34, 35: Hargreaves Associates et al., Orange County Great Park proposal, productive landscape (top); and water, riparian, and activity systems (bottom).

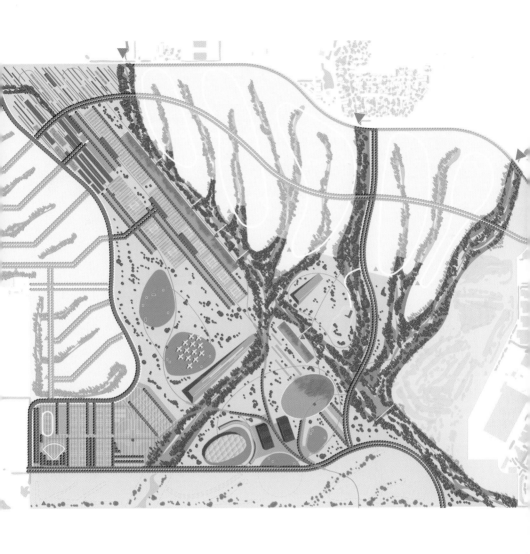

FIG. 36: Hargreaves Associates et al., Orange County Great Park proposal, speculative master plan.

NOTES

1. Martin Scheffer, Frances Westley, William A. Brock, and Milena Holmgren, "Dynamic Interaction of Societies and Ecosystems—Linking Theories from Ecology, Economy, and Sociology," in Lance Gunderson and C. S. Holling, eds., *Panarchy: Understanding Transformations in Human and Natural Systems* (Washington, D.C.: Island Press, 2001), 202.

2. Steward T. A. Pickett, Mary Cadenasso, and J. Morgan Grove, "Resilient Cities: Meaning, Models, and Metaphor for Integrating the Ecological, Socio-economic, and Planning Realms," *Landscape and Urban Planning* 69 (2004), 331.

3. C. S. Holling, D. W. Schindler, Brian W. Walker, and Jonathan Roughgarden, "Biodiversity in the Functioning of Ecosystems: An Ecological Synthesis," in Charles Perrings, Karl-Goran Maler, et al., eds., *Biodiversity Loss: Economic and Ecological Issues* (Cambridge: Cambridge University Press, 1995), 50.

4. Nina-Marie Lister, conversation with the author, January 2006. I thank her for this and other conversations that helped me grasp this ecological concept.

5. This language comes from Pickett, Cadenasso, and Grove, "Resilient Cities," 49.

6. I thank James Corner for my introduction to the possibilities for resilience within the design disciplines. My development of it as a topic is in no small part enabled through his own design work.

7. Galen Cranz, *The Politics of Park Design: A History of Urban Parks in America* (Cambridge, MA: MIT Press, 1982), 240.

8. Paul E. Cohen, *Manhattan in Maps, 1527–1995* (New York: Rizzoli, 1997), 100–05.

9. Elizabeth K. Meyer, "Scratching, Marking, Scarring the Surface: Large Park Design as an Act of Remembering and Forgetting Site Stories," (Large Parks: New Perspectives, conference held at the Harvard University Graduate School of Design, 2003).

10. For a discussion of Olmsted's theory of nature's effects, see Julia Czerniak, "Susceptible Bodies: The Absence of Distraction in Olmsted's Central Park," (Proceedings: 1996 ACSA Northeast Regional Conference, Buffalo, 1996.)

11. Edward Soja, "Exopolis: The Restructuring of Urban Form," in *Postmetropolis: Critical Studies of Cities and Regions* (Oxford: Blackwell Publishers, 2000), 233–63.

12. Ibid., 235.

13. Ibid., 238.

14. Ibid., 243. Soja cites Joel Garreau, *Edge City: Life on the New Frontier* (New York: Doubleday, 1991).

15. Rem Koolhaas, *SMLXL* (New York: Monacelli Press, 1995), 835.

16. Bruce Mau, *Lifestyle* (London: Phaidon Press, 2000), 288.

17. Sam Bass Warner, Jr., "Public Park Inventions: Past and Future," in Deborah Karasov and Steve Waryan, eds., *The Once and Future Park* (New York: Princeton Architectural Press, 1993), 17.

18. Wenche E. Dramstad, James D. Olson, and Richard T. T. Forman, *Landscape Ecology Principals in Landscape Architecture and Land-Use Planning* (Cambridge, MA, and Washington, DC: Harvard University Graduate School of Design and Island Press, 1996), 32.

19. Robert Glover, "City Making and the Making of Downsview Park," in Julia Czerniak, ed., *Downsview Park Toronto* (Munich and Cambridge, MA: Prestel and the Harvard University Graduate School of Design, 2001), 37–38.

20. See John Beardsley's essay, "Conflict and Erosion: The Contemporary Public Life of Large Parks," in this volume.

21. James Corner, conference presentation (Landscape Urbanism, Graham Foundation, Chicago, April 25–27, 1997). This attitude was invoked by noted environmentalist Ian McHarg in a study for Staten Island thirty years earlier. McHarg called Manhattan "a smear of low grade urban tissue" and "a great zoo, to which the animals make their way daily...and then return to the wild." He separated Manhattan "the city" from Staten Island "the wild," and by extension, nature. See Ian McHarg, *Design with Nature* (Garden City, NY: Natural History Press, 1969).

22. Sanford Kwinter, "Soft Systems," in Brian Boigon, ed., *Culture Lab* (New York: Princeton Architectural Press, 1996), 207–28.

23. Field Operations, *Lifescape: Fresh Kills Reserve*, report prepared for the City of New York, December 2002.

24. James Corner, conversation with the author, April 2003.

25. Ibid.

26. Ibid. Additionally, Lifescape calls itself a "reserve," a term that distinguishes the project from a park. Reserves are tracts of land set aside to preserve and evolve through management, whereas parks are recognizably "designed."

27. Field Operations, *Lifescape*.

28. Ibid.

29. Ibid.

30. Corner, conversation with the author, April 2003.

31. Nina-Marie Lister has explained that from a contemporary systems logic, we know that resilience and adaptability are tightly coupled, but as designers we are challenged to design for resilience in the hope that some adaptability will be achieved.

32. Field Operations, *Lifescape*.

33. Competition formula as listed in Czerniak, ed., *Downsview Park Toronto*, 75.

34. Christopher Hume, "If Downsview Park Matters, Why Has Nothing Begun?" *Toronto Star*, June 6, 2006.

35. The conceptual master plan is described by a 5:3:2 logic: five core values, three core zones, and two design strategies. The five core values of designing maintenance, designing education effects, redefining leisure, building a living database, and designing an icon are expressed in three core zones: the action zone, the promenade, and the cultivation campus. Each zone is then activated through two design strategies of "big moves" and "inventions," which emerge as the laundry list of programs one sees on the plan. Everything from the world's most "amazingly brilliant swing set" to "the millennium clock" promises to be reconciled within 600 acres of sustainability. Still, the absence of ecological and cultural details, of topographic and organizational specificity—the precise structural bones of the scheme—undermines the stated intention to guide the process of ecological self-organization. The park's ecological role in the city remains a promise, with suggestions of future connections to the ravine system without the infrastructure to enable it. These frameworks are essential underlays for the subsequent design development of the park.

36. David Anselmi, the vice president for park development, has reiterated the park's economic and ecologic goals, suggesting "great obligations" between commercial and civic desires, a statement that echoes Mau's notion of commercial tools working for civic ends. So how do Tree City's ideas go beyond the work of the logo? How, as the developer puts it, might the civic and the commercial add value to each other? The term "civic" here implies environmental sustainability, which development cannot compromise. Shortly after the competition, the sponsors conceptualized unconventional owner/tenant relations that they hoped would attract sustainable practices, recognizing, as they say, that "a good environment = good business & good business = a good park." Additionally, the development agency tried to negotiate using private space—such as the onsite movie studios—for public ends, a sort of private/public erasure that allows these spaces to be experienced as part of the park program. The concept of the commercial working for the civic, then, is a little more developed. First, marketing tools sell the initial scheme to the jury and the public, generating support. Second, new sorts of owner/tenant relationships were strategized in development details.

37. Anita Berrizbeitia, "Scales of Undecidability," in Czerniak, ed., *Downsview Park Toronto*, 116–25. See also Julia Czerniak, "Looking Back at Landscape Urbanism," in Charles Waldheim, ed., *The Landscape Urbanism Reader* (New York: Princeton Architectural Press, 2006), 105–23, for further discussion.

38. In Blenheim, the trees' understory is maintained as clean, straight lines by sheep who reach up to graze on low-hanging foliage.

39. Clare Lyster, Julie Flohr, and Cecilia Benites, "Assembled Ecologies: Infrastructure à la Carte," in *Twenty-first-Century Lakefront Park Competition* (Chicago: Graham Foundation, 2004).

40. James Corner, conversation with the author, July 2006.

41. Hargreaves Associates and Morphosis, *Orange County Great Park*, concept plan, September 2005.

42. Cranz, *The Politics of Park Design*.

43. Alan Tate, *Great City Parks* (New York: Spon Press, 2001).

44. Cranz, *The Politics of Park Design*.

JOHN BEARDSLEY has written extensively on public and environmental art, including the books *Earthworks and Beyond: Contemporary Art in the Landscape* (fourth edition, 2006) and *Gardens of Revelation: Environments by Visionary Artists* (1995). He teaches courses in landscape architectural history, theory, and writing at the Harvard University Graduate School of Design.

ANITA BERRIZBEITIA is author of *Roberto Burle Marx in Caracas: Parque del Este, 1956–1961* (2004) and coauthor, with Linda Pollak, of *Inside Outside: Between Architecture and Landscape* (1999). Her research focuses on the productive aspects of landscapes, particularly those of modern and contemporary landscape architecture. She is an associate professor of landscape architecture at the University of Pennsylvania.

JAMES CORNER is a registered landscape architect and urban designer, and founder and director of Field Operations. His design work has been recognized with numerous awards and is complemented by a body of writing on landscape architectural design and urbanism. He is author, with Alex MacLean, of *Taking Measures Across the American Landscape* (1996) and editor of *Recovering Landscape: Essays in Contemporary Landscape Architecture* (1999), which focuses on the revitalization of landscape architecture as a creative cultural practice. He is chair and professor in the Department of Landscape Architecture and Regional Planning at the University of Pennsylvania School of Design.

JULIA CZERNIAK is editor of *Downsview Park Toronto* (2001), which addresses contemporary design approaches to public parks and the relationship between landscape and cities. She is a registered landscape architect and founder and principal of CLEAR, a transdisciplinary design collaborative between architects and others that aspires to both strengthen its disciplinary identity and expand its range of operations. She is an associate professor of architecture at Syracuse University, where she teaches architecture studios as well as seminars on landscape theory and criticism.

GEORGE HARGREAVES is the Peter Louis Hornbeck Professor in Practice of Landscape Architecture at the Harvard University Graduate School of Design, where he was chairman of the Department of Landscape Architecture from 1996 to 2003. He founded Hargreaves Associates in 1982 and continues as design director. The firm's work has been published nationally and internationally and has won numerous design awards.

NINA-MARIE LISTER is editor, together with David Waltner-Toews and the late James Kay, of *The Ecosystem Approach: Complexity, Uncertainty, and Managing for Sustainability* (2007). A registered professional planner, Lister maintains an active practice with ZAS Architects. Her research, teaching, and practice focus on the confluence of landscape and ecological processes within contemporary metropolitan regions. Her research appears in various scholarly and professional practice journals as well as academic collections. She is an associate professor of urban and regional planning at Ryerson University in Toronto.

ELIZABETH K. MEYER has written numerous critical articles and essays on modern landscape architecture. She is currently working on a book entitled *Groundwork: Theoretical Practices of/in Modern Landscape Architecture.* She is an associate professor in the University of Virginia's Department of Architecture and Landscape Architecture.

LINDA POLLAK is coauthor, with Anita Berrizbeitia, of *Inside Outside: Between Architecture and Landscape* (1999) and a principal in the New York firm Marpillero Pollak Architects. Pollak has received grants from the NEA, the Graham Foundation, and the Milton Fund of Harvard University for her research on the relationships between architecture, landscape, and the city, which has appeared in numerous journals.

PAGES 2–3, 6, 15: photos by George Hargreaves

Foreword
Courtesy of James Corner/Field Operations; photo by Ellen Nieses

Introduction: Speculating on Size
FIG. 1: drawing by Ken Smith Landscape Architect
FIG. 2: courtesy of Ken Smith Landscape Architect; Jum Kim, illustrator
FIGS. 3, 4: drawings by West 8
FIGS. 5, 6: drawings by Hargreaves Associates
FIG. 7: drawing by Kertis Wetherby
FIGS. 8, 10: Julia Czerniak. drawings by Bruce Davison
FIG. 9: collage mapping by Ananda Kantner
FIG. 11: drawing courtesy of Richard Forman
FIG. 12: drawing by Bernard Tschumi Architects

Sustainable Large Parks: Ecological Design or Designer Ecology?
FIGS. 1, 2, 12: photos by the author
FIGS. 3, 4: drawing courtesy of James Corner/Field Operations
FIGS. 5–8: Nina-Marie Lister. Drawings by Bruce Davison
FIG. 9: photo courtesy of City of Toronto
FIG. 10: courtesy of James Corner/Field Operations; photo by Ellen Nieses
FIG. 11: drawing by James Corner/Field Operations

Uncertain Parks: Disturbed Sites, Citizens, and Risk Society
FIGS. 1, 2: photocollages by Julie Bargmann, DIRT Studio
FIGS. 3–7, 9: photocollages by Brett Milligan
FIG. 8: photographs by Brett Milligan

Matrix Landscape: Construction of Identity in the Large Park
FIG. 1: reprinted from Elizabeth Barlow Rogers, *The Forests and Wetlands of New York City*, 1st ed. (New York: Little, Brown, 1971)
FIG. 2: photo courtesy of James Corner/Field Operations
FIG. 3: courtesy of New York City Department of City Planning (NYCDCP)
FIGS. 4, 10–11, 16–20, 22–23: drawings by James Corner/Field Operations
FIG. 5: courtesy of WNYC.org
FIG. 6: courtesy of Robert Feldman Gallery
FIG. 6: copyright Estate of Robert Smithson
FIG. 7: photo by Linda Pollak
FIG. 8: Carlton Watkins, courtesy of www.yosemite.ca.us website
FIG. 9: courtesy of Staten Island Institute of Arts and Sciences
FIG. 12: courtesy of Green Matrix website, www.greenmatrix.com
FIGS. 13, 14: courtesy of NYCDCP
FIG. 21: courtesy of the Landscape Architecture section of the Gardenvisit.com website

Large Parks: A Designer's Perspective

FIGS. 1,11: collage map and drawing by Jason Siebenmorgen, Large Parks: New Perspectives

FIGS. 2–10, 13, 15–17, 20, 22–25, 29–34, 36–43, 45, 47–50: photos by George Hargreaves

FIGS. 12, 14, 18: collage map and drawings by Katherine Anderson, Large Parks: New Perspectives

FIG. 19: photo courtesy of Ananda Kantner, Large Parks: New Perspectives

FIG. 21: drawing by Ananda Kantner, Large Parks: New Perspectives

FIGS. 26–28: collage, map, and drawings by Rebekah Sturges, Large Parks: New Perspectives

FIGS. 35, 44: drawing by Anna Horner, Large Parks: New Perspectives

FIG. 46: drawing by Gina Ford, Large Parks: New Perspectives

Re-placing Process

FIG. 1: drawing by Katherine Anderson, Large Parks: New Perspectives

FIGS. 2, 16: drawings by James Corner + Stan Allen

FIG. 3: drawing by Anna Horner, Large Parks: New Perspectives

FIG. 4: drawing by Gina Ford, Large Parks: New Perspectives

FIGS. 5, 6, 8–10: photos by Anita Berrizbeitia

FIG. 7: drawing by Rebekah Sturges, Large Parks: New Perspectives

FIGS. 11, 12: drawings courtesy of Roger Sherman

FIGS. 13, 14: drawings courtesy of Anu Mathur and Dilip da Cunha

FIGS. 15, 17: drawings by James Corner/ Field Operations

Conflict and Erosion: The Contemporary Public Life of Large Parks

FIG. 1: photo courtesy of Mario Schjetnan

FIG. 2: photo by John Beardsley

FIGS. 3–5: photos and drawings by Caroline Chen, Large Parks: New Perspectives

FIGS. 6–8: collage map and photos by Darren Sears, Large Parks: New Perspectives

FIG. 9: credit unknown

Legibility and Resilience

FIGS. 1, 7–14, 29–31: drawings by James Corner/Field Operations

FIG. 2: drawing by Bruce Davison; Central Park figure courtesy of James Corner/ Field Operations

FIG. 3: photo matrix by Julia Czerniak

FIG. 4: © Living Earth/Spaceshots, Inc.

FIG. 5: credit unknown

FIG. 6: photocollage by Heather Vasquez and Nicholas Rigos

FIG. 15: reprinted from *CASE: Downsview Park Toronto*, Julia Czerniak, ed.

FIG. 16: drawing by Bernard Tschumi Architects

FIG. 17: drawing by Kristina Hill

FIG. 18: drawing by OMA + Bruce Mau

FIG. 19: drawing courtesy of Parc Downsview Park, Inc.

FIG. 20: stills from "Downsview Park: The Future: A Virtual Tour," Bruce Mau Inc., from www.pdp.ca

FIG. 21: drawing courtesy of Parc Downsview Park, Inc.

FIG. 22, 23: photos by Julia Czerniak

FIG. 24: credit unknown

FIGS. 25–28: drawings by Clare Lyster, Julie Flohr, and Cecilia Benites

FIGS. 32–36: drawings by Hargreaves Associates